Douglas Graham

A Treatise on Massage, Theoretical and Practical

Its History, Mode of Application and Effects, Indications and Contra-Indications

Douglas Graham

A Treatise on Massage, Theoretical and Practical
Its History, Mode of Application and Effects, Indications and Contra-Indications

ISBN/EAN: 9783337812331

Printed in Europe, USA, Canada, Australia, Japan

Cover: Foto ©berggeist007 / pixelio.de

More available books at **www.hansebooks.com**

A TREATISE

ON

MASSAGE,

Theoretical and Practical;

ITS

HISTORY, MODE OF APPLICATION AND EFFECTS,

Indications and Contra-Indications,

WITH RESULTS IN OVER FIFTEEN HUNDRED CASES.

BY

DOUGLAS GRAHAM, M. D.,

FELLOW OF THE MASS. MEDICAL SOCIETY;

Member of the Alumni Association of Jefferson Medical College, of the American Medical Association, of the British Medical Association, etc.

SECOND EDITION,
REVISED AND ENLARGED.

NEW YORK:
J. H. VAIL & COMPANY.
1890.

PREFACE TO THE SECOND EDITION.

THIS work has been thoroughly revised; it is to be hoped, improved. The enlargement consists of numerous additions, many of these confirmatory of statements previously regarded as doubtful. Two new chapters have been added, one on Local Massage for Local Neurasthenia, the other on The Treatment of Scoliosis by Means of Massage. Interesting items, long lost sight of in old literature, about the successful employment of massage as a *dernier ressort* in the case of Mary, Queen of Scots, and in that of Pope Clement VIII., have been inserted for the first time in their proper places in the History of Massage.

Attention is called to the fact that the motor points which give the best contraction to faradization are the same that give the best contraction to percussion. I believe it is for the first time also that high ankles are mentioned as a predisposing cause to sprains. Much new and valuable information from our *confrères* in Europe on the uses of massage in affections of the ear; in scoliosis; in fractures near and into joints; and in affections of the abdominal organs, has been added in a highly concentrated form. It has been impossible to do full justice in the limits of this work to the able and scientific articles of Dr. George Hünerfauth in "The Treatment of Chronic Typhlitis and Perityphlitis by Massage," and to the monograph of Dr. Rubens Hirschberg of Odessa on "Massage of the Abdomen for Dilatation of the Stomach and Chronic Dyspepsia." I might have added more of my own cases and less of those of others, but I think that the combination of experience offered is the fairest and most instructive. I have entirely forgotten to mention in the proper place two cases of myelitis in the early stages, and one advanced case of Landry's paralysis that recovered to an unexpected extent under massage, so that they could attend to active business.

The reviewers of the first edition of this work, almost without exception, showed lamentable ignorance of the whole subject, and it was painfully evident that most of them had had no previous acquaintance with massage. Nevertheless the freedom with which the book has been quoted and stolen from on both sides of the Atlantic is highly complimentary to the author. It is to be hoped that future reviewers of books on massage will speak of their comparative merits, and thus show that they know what they are talking about.

Since the first edition of this book was published, pictures of the position of the hands in doing massage have greatly increased, but have not improved. It is so rare to find one that stands the test of economizing time, space and effort in the most efficacious, agreeable and easy manner, that when we do, we may fairly conclude that this is more by accident than by design.

20 ~~DUMONT STREET~~, BOSTON, MASS.,
SEPTEMBER 1, 1890.

PREFACE TO THE FIRST EDITION.

NUMEROUS articles and abstracts on the subject of massage are scattered throughout medical literature and are practically buried alive in it. The object of the present volume is to resuscitate the most valuable of these and to unite them with the author's own experience. In doing this it has seemed better to bring forth illustrative cases rather than the deductions from these alone.

As an artist feels in comparing his best work with nature, recognizing more than any other wherein he has failed, so must every one feel who attempts to portray the functions of the human body in health and disease and the effects of therapeutical agents upon them.

The author sincerely hopes he has avoided the appearance of taking the oft-attempted, and astonishingly successful short road to fame by telling all that has ever been known or heard of in connection with the subject, as if he were the sole originator and proprietor of the same, without regard to its source.

The invitation of the publishers to illustrate my methods of doing massage with woodcuts has been declined, for even instantaneous photography can give but a poor conception of motion which can be done much better by words. All the illustrations of the methods of applying the hands in massage (save one plate by Pagenstecher, of Wiesbaden) show flagrant disregard for common-sense principles of economizing time, space, effort, and motion. It is hoped that the principles of massage are so clearly set forth in the following pages that they may be easily understood and made available by any one who has sufficient knowledge of anatomy, and aquaintance with natural and morbid consistency of tissues. With this knowledge pictures are unnecessary ; without it, they would be useless.

The good-natured liberality with which eminent neurologists often give preference to massage, rather than to electricity in which they are experts, is testimony of the highest order to the efficacy of manipulation, involving as it frequently does less necessity for their own valuable services; and the same may be said of many other physicians.

NOVEMBER, 1884.

CONTENTS.

A

PRACTICAL TREATISE

ON

MASSAGE.

CHAPTER I.

DEFINITION AND HISTORY OF MASSAGE.

"They be the best physicians which, being learned, incline to the traditions of experience; or being empirics incline to the methods of learning."
— *Bacon.*

Definition. — Massage, from the Greek μάσσω, I knead or handle ; Arabic, *mas'h*,[1] press softly ; is a term now generally accepted by European and American physicians to signify a group of procedures which are best done with the hands, such as friction, kneading, manipulating, rolling, and percussion of the external tissues of the body in a variety of ways, either with a curative, palliative, or hygienic object in view. Its application should in many instances be combined with passive, resistive or assistive movements, and these are often spoken of as the Swedish movement cure. There is, however, an increasing tendency on the part of scientific men to have the word massage embrace all these varied forms of manual therapeutics, for the reason that the word "cure," attached to any form of treatment whatsoever, cannot always be applicable, inasmuch as there are many maladies that preclude the possibility

---

[1] Hence those who inadvertently or through ignorance call *masseurs* "mashers" are oftentimes not, so far out of the way after all.

of recovery and yet admit of amelioration. Hence the word cure may lead people to expect too much; and on the other hand, the use of the word *rubbing* in place of massage tends to undervalue the application and benefit of the latter, for it is but natural to suppose that all kinds of rubbing are alike, differing only in the amount of force used.

History. — However massage may rejuvenate those who submit to its influence, the wrinkles of time cannot be removed from its own ancient visage. It is true they are sometimes forgotten, for so little attention have they received that Prof. Th. Billroth, of Vienna, in 1875, and Dr. Wagner, of Friedburg, in 1876, stated that there were many physicians in Germany who had never even heard of massage, and that it was then an everyday question as to what it meant, some even supposing that Dr. Mezger, of Amsterdam, was the originator of it.[1] Prof. Billroth continues by saying, " I can only agree with my colleagues Langenbeck and Esmarch that massage in suitable cases deserves more attention than has fallen to its lot in the course of the past ten years in Germany. My old experienced surgical assistant has already obtained a series of results both favorable and surprising and far exceeding my expectations of this method of treatment." And yet for sixty years prior to this the word massage had found a place in the medical literature of France, and valuable articles had from time to time appeared on the subject. Prof. Billroth says that massage is as old as surgery altogether. It being impossible to find out which is the older, perhaps, it would be more exact to say that massage is as old as mankind. Its origin has been well spoken of as lost in the night of time, and its use as hoary with antiquity. History informs us that massage has been partly practised from the most ancient times, amongst savage and civilized nations, in some form of rubbing, anointing, kneading, percussing, passive or mixed movements. Nor do we need to search far in order to account for this; for almost every one, when suddenly seized with a violent pain, instinctively and involuntarily seizes the

---

[1] Wiener Med. Wochenschrift, No. 45, 1875; Berliner Klin. Wochenschrift, Nos. 6 and 12, 1876.

painful place and attempts to relieve it by pressure or rubbing or both together, usually with the result of subduing the morbid and over-excited action of the nerves and of preventing blood stasis and effusion.

From what follows it will be observed that those who have thought it worth while to record their appreciation of massage have, in almost every instance, been men of note, eminent as physicians or philosophers, poets or historians, "who have left their footprints on the sands of time," from the days of Homer and Hippocrates down to these of Weir Mitchell and Billroth. Homer, about 1000 B. C., in the Odyssey tells us that beautiful women rubbed and anointed war-worn heroes, to rest and refresh them. In another part of the Odyssey (l. iii. v. 446) we read: "Meanwhile she bathed Telemachus, even fair Polycaste, the youngest daughter of Nestor. And after she had bathed him and anointed him with olive oil, and cast about him a goodly mantle, he came forth from the bath in fashion like the deathless gods." And again (lxxiv. v. 364): "The Sicilian handmaid bathed high-hearted Laertes and anointed him with olive oil, and cast a fair mantle about him." Such kindly marks of attention, we trust, were more for precept than example. Odysseus was more modest in accepting such hospitalities. "Then goodly Odysseus spake among the maidens, saying: 'I pray you stand thus apart, while I myself wash the brine from my shoulders, and anoint me with olive oil, but in your sight I will not bathe, for I am ashamed to make me naked in the company of fair-tressed maidens.'"

Among the old Greeks and Romans, massage in some primitive form or other was extensively patronized by people of widely different classes, from the patricians, the wealthy, and the learned downwards, to poor, decrepit old slaves; and for the most diverse purposes; with some as a means of hastening tedious convalescence, with others as a luxury in conjunction with the baths, and with others still to render their tissues supple and enduring preparatory to undergoing severe tests of strength, so that strains and ruptures would be less likely to occur. It was also used after the exercises and struggles, es-

pecially by the gladiators, in order to stroke away the ecchy-
moses and to relieve the pains of the bruises as well as to
reinvigorate them. Those who applied the rubbing and anointing were as different in character and qualifications as those
who received it. Sometimes it was done by medical practitioners themselves, sometimes by priests, at others by slaves,
but probably more often by those called *aliptæ* (from *alipes*,
swift of foot, nimble), whose business it was to anoint the wres-
tlers before and after they exercised, and who took care to keep
them sound and in good complexion.

In Athens and Sparta, the gymnasium was a state institution
frequented by any free-born citizen. Solon, the Athenian law-
giver, 638–559 B. C., watched over it with jealous care and pun-
ished with death any slave detected within its sacred precincts.
The gymnastics of the ancients were divided into athletic, military, and medical. Herodicus, one of the masters of Hippocrates, in the fifth century B. C., first proposed gymnastics for
the cure of disease and the preservation of health. To such an
extent did he carry his ideas that he compelled his patients to
exercise and to have their bodies rubbed, and by this method
he had the good fortune to lengthen for several years the lives
of so many enfeebled persons that Plato reproached him for
protracting that existence of which they would have less and
less enjoyment.[1] Herodicus was not only a physician, but a
pædotribe, *i. e.*, the superior officer acquainted with all the
prescribed movements in the ancient gymnastics and who carried
them into effect. He was induced to study gymnastics from a
medical point of view, from having been benefited by them, and
unlike many physicians, by thus applying his own practice to
himself, it is said that he cured himself of bodily weakness and
ill health, and attained the age of 100 years.[2]

Herodotus, 484 B. C., called by Cicero the Father of History,
says that there were specialists in Egypt, a particular physician
for each disease. "The art of medicine is thus divided amongst

---

[1] Hufeland's Art of Prolonging Life.
[2] Allgemeine Orthopädie, Gymnastik und Massage, von Prof. Dr. Friedrich Busch,
Berlin.

them ; each physician applies himself to one disease only, and not more. All places abound in physicians; some physicians are for the eyes, others for the head, others for the teeth, others for the parts about the belly, and others for internal diseases." In a passage from Herodotus we are informed that after having poured upon the body a greasy mixture, each part ought to be rubbed, passing the hands from above downwards. At the commencement the friction ought to be gentle and slow, then it should become rapid and accompanied with pressure, whilst towards the end the friction will again become gentle.

The writings of Plato abound with references, direct and in-direct, to friction. His teacher, Socrates, 470-399 B. C., the Greek philosopher and representative of the finest Athenian culture, asserts that the first country which gave birth to men was Attica; and in proof of this he adduces the fact that what-ever brings forth offspring is provided by nature with food for that offspring. Now Attica, he says, is the native soil of wheat and barley by which the race of men is most excellently and best nourished. And after this she caused to grow for her children olive oil, the assuager of pain. Here Socrates, by the mouth of Plato, esteems oil as only less necessary to human life than wheat and barley, referring to its use in the way of friction which often does allay pain in a remarkable manner; but taken internally it has no such effect.[1] Pliny most dis-tinctly says that amongst the errors which the Greeks imported into the gymnasium was the use of oil, which Anacharsis, a philosophic Scythian of the 6th century, B. C., considered to be the medicine of madness, because the athletes seemed to be maddened by it. By the injudicious use of oil, I have seen patients made very angry, but not quite "mad." Nearly every substance capable of being rubbed on the human body has had wonderful curative virtues ascribed to it. Many of these, besides being intensely disagreeable, are doubtless worth-less, and for the resulting benefit we must look for something that is common to them all, and this we find in their method of application, namely, the rubbing.

---

[1] The Anatriptic Art, pp. 3 and 4.

The wisdom of the ancients appears to great advantage in some of their remarks about rubbing, and it requires years of practical acquaintance with massage in order to fully appreciate them. The aphorisms of Hippocrates, 460 to 380 B. C., on this subject embodied the wisdom of the past and presaged the development of the future to a greater extent than either ancient or modern writers on massage have shown any evidence of understanding. "The physician must be experienced in many things," says Hippocrates, "but assuredly also in rubbing; for things that have the same name have not always the same effects. For rubbing can bind a joint that is too loose and loosen a joint that is too rigid."[1] And again, "rubbing can bind and loosen; can make flesh and cause parts to waste. Hard rubbing binds; soft rubbing loosens; much rubbing causes parts to waste; moderate rubbing makes them grow." This is the earliest definite information about massage, and its truth and meaning are fully realized when the necessary previous conditions exist. Hippocrates was wiser than he was aware of, as we learn from the word that he used to designate the process of rubbing, namely, *anatripsis*, literally the art of rubbing up, and not down. In those days the circulation of the blood was not understood, the ancients supposing the arteries to be filled with air, and hence their name, which still clings to them. Though Galen more than 500 years later won the honor of pointing out the fact that the arteries contain blood in the living body, but are for the most part empty after death; yet the circulation of the blood was not understood until discovered and demonstrated in the year of our Lord 1628 by the immortal Harvey, who lost his practice in consequence. The observations of Hippocrates must have been very accurate to discern that rubbing upwards in the case of the limbs had a more favorable effect than rubbing downwards, and doubtless in this manner he had experience in promoting the resorption of effusions; for it is now well known that upward friction on the limbs favors the return of the circulation, relieves blood stasis, and makes more room in the veins and lymphatics for the carrying away

---

[1] Hippocrates. " Peri Arthron." Littré, Vol. iv., p. 100.

of morbid products. This affords an illustration of "science following art with limping pace," which so frequently happens in the practice of medicine.

By appropriate massage, passive and resistive movements, atrophied muscles, tendons, and ligaments would have their circulation accelerated and increased, and consequently their nutrition and innervation improved so that they would become larger and firmer, thus binding closer a joint too lax and making it stronger. By the same means involuntary tension of the muscles, adhesions, effusions, and hyperplastic tissue may be removed, so that a joint stiff from such causes would become more flexible. Therefore the saying of Hippocrates, that anatripsis will bind closer a joint that is too lax and relax a joint that is too rigid, is not so paradoxical as it seems. These remarks also in part refer to the fact that " rubbing can make flesh and cause parts to waste," in its local application ; but in its general application the same effects have been observed and much more fully referred to by Dr. S. Weir Mitchell in " Fat and Blood, and How to Make Them." People who have a normal quantity of adipose tissue sometimes lose much of it, to their detriment, by the excessive use of massage. But even this feature can sometimes be utilized to advantage in cases where fat is superabundant, soft and flabby, with a want of tone and tension in the areolar tissue, and in these it will be found that hard rubbing binds. " Soft rubbing loosens" not only abnormally tough and *matted* conditions of skin and superficial fascia, but also involuntary tension of muscles, both of which conditions, if looked for, may often be found generally as well as locally in overtaxed and debilitated people. Such a state of these tissues would often seem to be a physical expression of too great mental tension which the patient, like his muscles, is unable to relax. And here comes the necessity of a careful discrimination ; for, if a patient whose condition corresponds to that first described should receive such vigorous rubbing as often passes for massage in these days, and the vigor of which would really seem to be necessary to relax the tenseness of the tissues, the trouble would in all probability be aggravated, for reflex action and consequently still greater tension would be excited by the pressure of rough friction and manipulation upon terminal nerve filaments which are already in a state of irritation. Though it does not appertain to the history of massage, yet it may not be amiss to say here that an admirable preliminary measure in such cases is a warm bath, which is grateful and soothing to the patient, solicits the blood to the surface, softens the cuticle and removes the epithelial débris, and also relaxes the skin and to some extent the tissues beneath it. "Moderate rubbing makes parts grow" implies that the tissues to be rubbed are insufficiently nourished, and that if they be immoderately rubbed, their vitality will be lessened, their natural nervous irritability exhausted and a state of congestion induced highly unfavorable to their proper nutrition.

These brief sayings of Hippocrates on anatripsis serve partly to show at the same time why he was considered a man of transcendental genius and justly styled the "Father of Medicine," who, having raised the art from a system of superstitious rites practised wholly by the priests, to the dignity of a learned profession, was then accused by his jealous contemporaries of having made too free use of the writings of others, and of having burned the collection to conceal his plagiarisms. It was supposed that he had ample opportunity to do this in his capacity as librarian of the famous medical school of Cos, of which he was also chief.

Asclepiades, 128–56 B. C., a celebrated Greek physician, founded a school, practised at Rome, and was very popular with the Romans on account of his simple and agreeable remedies. He thought that physicians ought to cure their patients safely, speedily, and pleasantly; and he relied mainly on diet, bathing, exercise, and friction. His representation of treatment by motion found great respect. He considered that the body was composed of innumerable canals, endowed with sensation and regularly distributed, in which moved the nutritive juices and plastic atoms. So long as this movement went on without disturbance, health continued; on the contrary, disturbance caused sickness. The normal movement of the juices would be disturbed by abnormal increase of these atoms, by their irregular distribution, by too great blending together, and by two swift motion of the same, and also by constriction and dilation of their canals. Proceeding upon these principles, Asclepiades renounced almost entirely the use of medicine and attempted to restore free movement of the nutritive fluids and atoms by means of rubbing ; one use of which amongst others he particularly recognized was that gentle stroking had a soporific influence. With this he also combined active and passive motion.

Cicero, the Roman orator, philosopher, and statesman, B. C. 106–43, considered that he owed as much of his health to his anointer as he did to his physician. At the beginning of his career he was stopped by his infirm health and for its restoration he travelled in Greece and Asia. In time he overcame his violent manner of speaking which his feeble frame was unable to bear, and he met his death by assassination with more bravery than he had anticipated; all of which would indicate a marked improvement in his general health, and from this, better control of his mind and nervous system, due in great part, we may certainly suppose, to the tonic and sedative effects of rubbing. Plutarch tells us that Julius Cæsar, B. C. 100–44, had himself pinched all over daily as a means of getting rid of a general neuralgia. The distinguished Roman physician Celsus, who flourished about the commencement of the Christian era, spoke wisely and well about rubbing, in saying that it "should sometimes be applied to the whole body, as when an invalid requires his system to be replenished." And again: "Chronic pains of the head are relieved by rubbing the head itself. But far more frequently when one part is in pain another must be rubbed, particularly when we desire to *draw matter* from the upper or middle part of the body and therefore rub the extremities. A paralyzed limb is strengthened by being rubbed. If certain limbs only are rubbed, long and powerful rubbing may be used, for the whole body cannot soon be weakened through a part. But when weakness of the body needs this cure over its whole extent, it ought to be shorter and more gentle than local rubbing, so as only to soften the superficial skin that it may be enabled the more easily to receive new matter from the food. A thing becomes constricted when we take away that which by its interposition produced relaxation, and softened when we remove that which caused its hardness, and filled, not by the rubbing, but by the food which afterward penetrates to the skin which has been relaxed by a kind of digestion or removal of its tissue." For the purpose of dispersing local deposits, and thus relieving the pains occasioned thereby, Celsus says that "one must use friction also, particularly in the sun, and several times daily, in order that the matters which by their collection have produced the mischief, may be the more easily dispersed." The mistake with Celsus was that he advised friction for almost every disease, and sometimes contradicted himself, but not altogether without reason, as the following sentences show: "As rubbing is rightly applied after the cessation of an illness, so it must never be used during the increment of a fever, but if possible when the body shall have been wholly free from it. A patient is in a bad state when the exterior of the body is cold, the interior hot with thirst: but, indeed, also the only safeguard lies in rubbing, and if it shall have called forth the heat into the skin it may make room for some medicinal treatment."

The wise and able Emperor Hadrian, A. D. 76–138, who will be so well remembered as having built the wall from the Solway Frith to the Tyne, and whose reign was distinguished by peace and beneficent energy, one day seeing a veteran soldier rubbing himself against the marble at the public baths, asked him why he did so. The veteran answered, "I have no slave to rub me;" whereupon the Emperor gave him two slaves and sufficient to maintain them. Another day several old men rubbed themselves against the wall in the emperor's presence, hoping for similar good fortune, when the shrewd Hadrian, perceiving their object, directed them to rub one another!

The health of the celebrated Roman advocate Pliny, which was never very strong, had been shaken by a severe illness the preceding year, A. D. 102. His life, he tells the emperor in one of his letters, had been in danger. He availed himself of a mode of treatment which, it is presumed, was much in vogue at that time. He procured the services of a medical practitioner who cured many of his patients by the process of rubbing and anointing, and so much benefit did he derive from the remedy that he asked the emperor to grant the physician, who was either a Jew or a Greek, the freedom of the city and the privileges of Roman citizenship.

The art of embellishing was very much cultivated amongst the ancients, and for this purpose physicians did not disdain to make use of the palette, an ovoid disc terminating in a handle, with which percussion was done. For this reason Pliny likened them to schoolmasters. In the principal cities there were establishments to which slaves having some slight deformity were taken at the expense of their masters in order to undergo a course of treatment for the purpose of deceiving their buyers and to acquire the comeliness of figure which they lacked. The procedure was percussion by means of the palette. These places were considered to be of bad reputation, and sometimes women went to them secretly to seek for that freshness and rotundity in which they were deficient, and their weakness yielding to their vanity, they endured the blows of the palette which was necessary to use at a great rate, and this was only interrupted by palpation, contrectation, and all the resources of *psellaphie*, a word supposed to be synonymous with massage.[1] Men used up by excess also resorted to these places for the rapidly revivifying effects of such treatment.

Martialis, of eminent literary fame, about the year of our Lord 100, refers to manipulation in the following words:

"Percurrit agile corpus arte tractatrix
Manumque doctam spargit omnibus membris."

Galen, A. D. 130–200, the most learned physician and the most accomplished man of his age, whose authority in medical matters was regarded in Europe as almost supreme for a thou-

---

[1] Dict. des Sciences Médicales, 1819.

sand years, recommended friction in a great number of diseases, generally as auxiliary to other means. At Pergamus, his native city, he was appointed physician to the school of gladiators. He was deeply interested in exercises and friction, and laid down minute directions concerning the latter, part of which it would be well to remember at the present day. "If any one," says he, "immediately after undressing proceed to the more violent movements before he has softened the whole body, and thinned the excretions, and opened the pores, he incurs the danger of breaking or spraining some of the solid parts. There is danger also of the excretions, in the rush of moving spirits, blocking up the pores. But if beforehand you gradually warm and soften the solids and thin the fluids, and expand the pores, the person exercising will run no danger of breaking any part, nor of blocking up the pores. Hence, in order to insure this result, it is proper by moderate rubbing with a linen cloth to warm the whole body beforehand, and then to rub with oil. For I do not counsel the immediate application of the grease before the skin is warmed and the pores expanded, and, generally speaking, before the body is prepared to receive the oil; and this will be accomplished by a very few turns of the hands, without pain and moderately quick, having in view to warm the body without compressing it; for you will perceive while this is being done a blooming redness running over the skin; and this is the time to apply the grease to it, and rub with bare hands, observing a medium hardness and softness in order that the body may not be contracted and compressed, nor loosened and relaxed beyond the fitting extent, but be kept in its natural state. And one should at first rub quickly, and afterwards, gradually increasing it, push the strength of the friction so far as evidently to compress the flesh, but not to bruise it. . . . In using friction preparatory to the gymnastic exercises, the use of which is to soften the body, the middle quality between hard and soft should prevail, and all else should take its fashion accordingly. . . . And I recommend the imposition and circumflexion of the hands to be varied, in order that all the fibres of the muscles, as completely as possible, in every part may be rubbed; for the opinion that transverse rubbing, which some call circular rubbing, hardens and condenses, and contracts and binds the body, but that perpendicular rubbing rarefies and dilates and softens and unbinds, is a mark of the same ignorance from which proceed most of the other assertions made by gymnastic professors on the subject of rubbing."

Sudden and violent efforts at running, jumping, lifting, and the like, by those unaccustomed to them, especially if they have passed the meridian of life, are apt to cause rupture and strain of muscular and tendinous fibres, owing to a lack of suppleness in these tissues. It would be difficult to improve on the preventive treatment of such injuries advised by Galen. The multitude of modern greasers and bruisers, who are supposed to be doing massage, might profit by the hint of Galen as to the best time of applying unctuous materials ; and those who attempt to explain their doings by calling tendons, nerves, and other such nonsense had better take heed to the criticisms of Galen on the assertions of gymnastic professors. No wonder that he took so much interest in exercise and kindred measures for the improvement and maintenance of health, for history tells us that till the age of thirty years he was weakly, but became strong and of good health by devoting several hours a day to bodily exercise, and in this way cured a host of sicknesses and weaknesses in others.

It was sometimes amusing and absurd to what an extent some of these old writers noted their appreciation of rubbing, especially in the care of that canine, domestic idol which too often takes the place of the baby. Arrian, who probably lived about the year of our Lord 243, says: "And great is the advantage of rubbing to the dog of the whole body — not less than to the horse, for it is good to knit and strengthen the limbs, and it makes the hair soft and its hue glossy, and it cleanses the impurities of the skin. One should rub the back and the loins with the right hand, placing the left under the belly, in order that the dog may not be hurt from being squeezed from above into a crouching position ; and the ribs should be rubbed with both hands ; and the buttocks as far as the extremities of the feet; and the shoulder blades as well. And when they seem to have had enough, lift her up by the tail, and having given her a stretching let her go. And she will shake herself when let go, and show that she liked the treatment." — [Arrian Cynegeticus.]

Aside from the beneficial effects of the friction, the stretching also is of value, for we know that people in health, as well as animals, stretch before and after sleep, and this is partly involuntary. The influence of alternately stretching and relaxing the fasciæ in favoring the flow of lymph has been most interestingly demonstrated outside of the living body, thus showing how one of the most important physiological processes is carried on, and the necessity of semi-involuntary movements taking place when rest or fatigue has kept a person too long in one position. But more of this hereafter in its proper place.

Oribasius, a Greek, who early acquired a high reputation, and was taken by the emperor Julian to Gaul as his physician, describes in wearisome detail the apotherapeia or perfect cure, meaning the last part of the ancient gymnastics which consisted of bathing, friction, and inunction for the purpose of obviating fatigue or curing disease. "The apotherapeia," says he, "has two objects, that of evacuating superfluities and of preserving the body from fatigue. The former is common to exercise considered as a whole, for we regard exercise as having two effects: that of strengthening the solid parts of the body and of evacuating the superfluities. The peculiar aim of apotherapeia is to combat and to prevent the fatigue which habitually follows immoderate exercise, and the nature of this aim will indicate to us how it is necessary to perform the apotherapeia; for if we propose to evacuate precisely the superfluities of the solid parts of the economy, which after having been warmed and attenuated by exercise still remain in the system, we must use friction by many hands and varied with rapidity, in order that as much as possible no part of the individual whom we rub be uncovered. During the friction we ought to extend the parts which we rub, and besides, we will prescribe what we call retention of the breath. . . . We are of opinion, that it is well to extend the parts which we rub, so as to evacuate through the skin all the superfluities which find themselves between it and the subjacent flesh; and it is for the same reason that an important part of the apotherapeia consists in holding the breath, which causes tension of all the muscles of the chest and relaxation of those of the belly and of the diaphragm; and thus the excrements will be pushed downwards," etc., etc. If, in place of relaxation, Oribasius had used the word contraction, the cause would then have agreed with the result, but the muscles and their actions were not so well understood in those days as now. What he considered superfluities we would doubtless call excretions.

" But all the physicians and philosophers of antiquity knew no better means of strengthening the vital principle and prolonging life than by moderation ; by the use of free and pure air and bathing, and above all by daily friction of the body and exercise. Rules and directions were laid down for giving gentle and violent motion to the body in a variety of ways, hence arose a particular art called the gymnastic ; and the greatest philosophers and men of learning never forgot that the body and soul ought to be exercised in due proportion. This art of suiting exercise to the different constitutions, situations, and wants of man ; of employing it above all as the means of keeping his internal nature in proper activity, and thereby not only rendering the causes of disease ineffectual, but also curing diseases which have already appeared, they indeed brought to an extraordinary degree of perfection."

It is astonishing how many there are who have thought that massage was indigenous to their own or some other country. Thus the pleasing statement that the manœuvres of massage have been imported into Europe from Syria, Palestine and the East, in consequence of the crusades against the Saracens, would lead us to think that massage had not been heard of before that period in Europe. But it is more likely that it had fallen into neglect and been forgotten.

CHAPTER II.

" Vivifying old and neglected truths needs as much prophetic insight as to
see near truths for the first time."

HAVING seen that massage has been practised after a fashion
by the Greeks and Romans, it is reasonable to suppose that it
was also made use of by their common ancestors, the Aryans.
It may not be amiss, just here, to let history tell us who the Aryans were, and where their original place of abode was. The
Aryan branch of the Caucasian race includes nearly all the past
and present nations of Europe, and it is that division to which
we ourselves belong. The Aryans were a fair-skinned, noble
people, progressive, practical, and warlike, and it is said that
they speedily subdued the country adjacent to them, and also
the peninsula of India, 3000 B. C. The original seat of the un-
divided Aryan stock was to the northeast of Persia, in the region of the Oxus and Jaxartes rivers. The water of the Oxus
is said to have been extremely soft, so as to have made the skin
of those who bathed in it to glisten. Now we certainly may be
allowed to imagine that from the pleasant sensation of drawing
the hand over the skin in admiration of the effect of the water
of the Oxus upon it may have arisen the methodic use of friction amongst the Aryans. It may be very interesting to learn
that some form of friction, kneading, or handling has been in
use amongst savage races from the most remote periods ; but
such testimony alone would not be of much value did we not
learn at the same time that such treatment had been known and
used, and its merits recognized and recorded, by the descendants of

that branch of the human race who have made the greatest progress in the history of the world, and who have contributed more than any other towards its civilization. We refer to the Aryans and their descendants. But the Greeks and Romans were not the only members of the Aryan family who practised rubbing in the early ages. Strabo tells us that the Indians contemporary with Alexander, 326 B.C., esteemed friction highly. " In the way of exercise," he says, " they think most highly of friction ; and they polish their bodies smooth with ebony staves and in other ways. The king while receiving foreign ambassadors listens and is rubbed at the same time." In India, as in ancient Greece, the groom rubs his horse with his own naked hand, in consequence of which it is said that their horses have a much finer coat than the English, which receive such attention from currycomb and brush. Amongst the hygienic principles laid down in the *Ayur-Veda* (Art of Life) in the early Sanscrit of the first century are these: One ought to rise early, bathe, wash the mouth, anoint the body, submit to friction and shampooing, and then exercise. The word shampooing is of Hindoo origin, and, however refreshing the process and applicable to the Turkish bath, it should not be used as a synonym for massage, which is a different, more graceful, and more effectual procedure.

Paracelsus, 1492–1541 A. D., professor of surgery at Bâle in 1526, a remarkable man, though often intoxicated and guilty of gross immoralities, in his " Liber de Vita Longa," extols the effects of friction on the human body as indispensable to health.

Ambroise Paré, 1517–1590 A. D., the most renowned surgeon of the sixteenth century, though not recognized by the faculty, as he was only a barber surgeon,[1] the inventor of the ligation of arteries which is the foundation of modern surgery, surgeon under four French kings, a devout Huguenot, but spared at the massacre of St. Bartholomew on account of his surgical skill,— good old Ambroise states in his works, which were published in 1575, that friction was in great esteem in his time. He describes three kinds of friction — gentle, medium, and vigorous — and the

---

[1] Prof. Gross narrates that when Ambroise Paré was a young man he lived with a noble family to do the shaving, the surgery, and to read the family prayers.

effects of each. In dislocations, he recommends that the joint should be moved about, this way and that way, not violently, but in order to resolve the effused fluids, and extend the fibres of the muscles and the ligaments, so as to facilitate the reduction. From this it is apparent that he knew the influence of passive motion in promoting absorption, the rationale of which has been so well studied by German physiologists.

Mercurialis, 1530–1606 A. D., an eminent Italian physician who graduated at Padua and later occupied a chair of medicine in that celebrated University, published in 1573 a treatise entitled "De Arte Gymnastica," in which he brings prominently forward the benefits to be derived from active, passive, and combined movements. Alpinus, 1553–1617 A. D., was a celebrated Italian botanist who occupied the chair in botany at Padua in 1593. In his "Medicina Ægyptia," Chapter XVIII., he says that frictions are so much in use amongst the Egyptians that no one retires from the bath without being rubbed. For this purpose the person is extended horizontally; then he is malaxated, manipulated, or kneaded, and pressed in divers manners upon the various parts of the body with the hands of the operator. Passive motion is then given to the different articulations. Not satisfied with *masséing*, flexing and extending the articulations alone, they exercise the same pressures and frictions upon all the muscles, the effect of which is thus described by Savary: " Perfectly *masséed*, one feels completely regenerated, a feeling of extreme comfort pervades the whole system, the chest expands, and we breathe with pleasure ; the blood circulates with ease, and we have a sensation as if freed from an enormous load ; we experience a suppleness and lightness till then unknown. It seems as if we truly lived for the first time. There is a lively feeling of existence which radiates to the extremities of the body, whilst the whole is given over to the most delightful sensations ; the mind takes cognizance of these, and enjoys the most agreeable thoughts ; the imagination wanders over the universe which it adorns, sees everywhere smiling pictures, everywhere the image of happiness. If life were only a succession of ideas, the rapidity with which memory retraces them, the vigor with which the mind runs over the extended chain of them, would make one believe that in the two hours of delicious calm which follow a great many years have passed."

Fabricius ab Aquapendente was a pupil of Gabriel Fallopius, and later professor of surgery at Padua, where he enjoyed a high reputation for many years, from 1565 onwards. Besides works on surgery, he was the author of a treatise, "De Motu Locali Secundum Totum," in which he again brought massage to honor. He most warmly recommended this treatment by rubbing, kneading, and scientific movements as a rational measure in joint affections.

Mary, Queen of Scots, was stricken down Oct. 7th, 1566, with a malignant, intermittent typhus fever, doubtless caused by fatigue and annoyance at the wretched conduct of her husband. She was very ill and sank rapidly. Convinced that her last hour had come, she calmly prepared for death. She forgave all who had in any way offended her, and craved pardon of all whom she had in the slightest aggrieved. Soon she became cold and rigid, her form straightened out, her pulse and respiration were no longer preceptible. All despaired of her, save her physician News, who, hoping against hope, continued to use violent frictions, and at length succeeded in restoring her to life. She then began rapidly to improve, but her death meantime had been reported in Edinburgh.[1]

Clement the Eighth, one of the greatest Popes that the Church has ever had, was a great sufferer from gout in his hands and feet. His friend, Saint Philip Neri, was very fond of him and visited him as often as he could, but was frequently prevented from doing so by sickness or other causes. It was about Easter, 1595, that the Pope had an unusually severe attack and was ordered by his physician to keep his bed. When Philip heard of this he had a great desire to relieve him. He first prayed for the Pope with great fervor and then went to visit him. When he came into the room Clement was in so much pain that he could not bear anyone to touch the bed he lay upon, and he begged Philip not to come near to him. But Philip

---

[1] Life of Mary, Queen of Scots, by Donald McLeod.

moved gently towards him and Clement again entreated that no one should touch him. With a smile of affectionate sympathy Philip replied: "I am not sorry for the gout, Holy Father, for that compels you to rest; but I am very sorry for the pain you suffer. Your Holiness need not fear; let me do as I please." And without another word he seized the suffering hand and pressed it with great affection. The pain immediately disappeared and Clement cried out: "Go on touching me, Father, it gives me the greatest relief." The Pope was thus healed, so it is said, and spoke of it as a miracle to the Cardinals for examining Bishops, and often adduced it as proof of Philip's sanctity. From that time forward, and even after Philip's death, whenever the Pope was suffering from gout, he commended himself to Philip and the pain was at once relieved.[1]

What seems remarkable to us is that one so delicate as Saint Philip should apparently exert such power of relief as we are led to suppose that he did in this case. Either the Pope did not commend himself to his Heavenly Father for relief, or if he did, evidently it had less effect in relieving him than commending himself to Saint Philip. That the Pope's faith was sufficient to relieve him after the death of the Saint surpasses all the mind-cures and faith-cures of the present day, for they require a living and active agent through which to act on their patients or dupes.

In the "Miroir de la Beauté" of Guyon, 1615, exercise and friction are advised, and it is considered necessary to have the body rubbed gently by some person who has soft hands.

The illustrious Sydenham, 1624-1689, abandoned the routine system of practice then prevalent, and based his own upon the theory that there is in nature a recuperative power which ought to be aided and not opposed. An example of this is found in his saying that, if any one knew of the virtues of friction and exercise, and could keep this knowledge secret, he might easily make a fortune. This is fully exemplified at the present day, for in every city of the United States, and indeed of the whole civilized world, there may be found individuals claiming

---

[1] Life of Saint Philip Neri, Vol. II.

mysterious and magical powers of curing disease, setting bones, and relieving pain by the immediate application of their hands. Some of these boldly assert that their art, or want of art, is a gift from Heaven, due to some unknown power which they call magnetism, while others designate it by some peculiar word ending with pathy or cure, and it is often astonishing how much credit they get for their supposed genius by many of the most learned people. Let a fisherman forsake his boat, or a blacksmith his anvil, or a carpenter his bench, or a shoemaker his shop, and proclaim that he has made the wonderful discovery that he is full of magnetism and can cure all diseases, and be he ever so ignorant and uncouth, he is likely to have in a remarkably short space of time a large clientèle of educated gentlemen and refined ladies. It is not meant to imply that the previous occupation of these people is at all to their discredit, but were they capable of giving a rational explanation of their doings, the halo of mystery would be removed from around them, and their prestige and patronage would suffer a sudden decline.

Hoffman, 1660–1742, who was physician to the King of Prussia, we are not likely to forget so long as the anodyne which still bears his name continues to be so useful. In his " Dissertationes Physico-Medicæ," 1708, he says that exercise is the best medicine for the body, and that we cannot imagine how salutary and favorable to health it is, for it excites the flow of the spirits, and facilitates the excretions from the blood. He extols the passive, active, and mixed movements of the ancients as well as the apotherapeia already referred to.

In the year 1698, Paullini offered to libertines, as if they were sufferers (which no doubt they were), what he considered a very efficacious remedy. It was nothing more nor less than flagellation, percussion, and slapping; and probably this was not administered with half the severity it ought to have been, in order to produce other than palliative results. Rubbing the back was used by the ancients for sterility. The Roman ladies allowed themselves to be whipped with strips of leather in such cases. (Ovid.)

In 1780, Simon André Tissot, professor of clinical medicine at the University of Pavia, interested himself in massage, and wrote an "Essai sur l'utilité du mouvement ou des différentes exercices du corps et du repos dans la cure des Maladies," or "Gymnastique Médicinale et Chirurgicale."

A curious old book is that entitled "A Full Account of the System of Friction as Adopted and Pursued with the greatest Success in Cases of Contracted Joints and Lameness from Various Causes, by the late eminent Surgeon, John Grosvenor, Esq., of Oxford. With Observations by William Cleobury, Member of the Royal College of Surgeons. Third Edition. With a Portrait and Memoir of the Author. Oxford, 1825." About a century ago, Mr. Grosvenor was professor of surgery for many years at Oxford, where his skill and reputation became so great that he was soon in possession of all the surgical practice at Oxford, and on every side of it within a radius of thirty miles. He was undoubtedly a man of ability, for, in addition to his extensive practice, he edited a newspaper during his breakfast hour, and rendered gratuitous services to the poor from eight to ten in the morning. He practised simply as a surgeon, and would not invade the province of a physician nor condescend to soil his fingers with the preparations of pharmacy. "In the latter period of his practice, Mr. Grosvenor rendered himself celebrated throughout the kingdom by the application of friction to lameness or imperfections of motion arising from stiff or diseased joints. He had first used it with success in a complaint of his own, a morbid affection of the knee, and by degrees its efficacy was so acknowledged that he was visited by patients from the most distant parts, of the highest rank and respectability, among others by Mr. Hey, the able surgeon of Leeds. Those who were benefited by the process pursued under his own immediate superintendence in cases of this sort, and from total inability have been restored to a free use of their limbs, were best able to attest his merits. That he was scarcely in any instance known to fail, was perhaps attributable to the circumstance that he used his utmost efforts to dissuade from coming to Oxford every one of whose case, from previous communications, he entertained any doubt. Possessed at this time of affluence, he became very indifferent about business, and at a time of life when he was still capable of active exertions and his strength was but little impaired, he began to contract his practice. For the last ten years of his life he had wholly given up his profession, except in the instances of his patients requiring friction." Mr. Grosvenor considered friction highly improper in all cases of inflammation, in scrofulous cases tending to suppuration, in cases of inflammatory gout and rheumatism, and useless in cases of true anchylosis. The cases in which he found this remedy most serviceable were contractions of the joints attended with languid circulation and thickening of the ligaments, in those cases in which there is too great secretion of the synovial fluid in the joints, after wounds in ligamentous, tendinous, or muscular parts when the function of the limb is impaired, in cases of paralysis, in cases of chorea combined with attention to the system, after violent strains of the joints, in incipient cases of white swelling, after fractures of the articulating extremities of the joints when stiffness remains after union, in cases of dislocation of the joint when the motion is impaired some time after reduction, and in weakly people where the circulation is languid. The observations of Mr. Grosvenor have been, in the main, confirmed by others, most of whom evidently consider their own experience unique and unprecedented.

To Peter Henrik Ling, poet and physiologist of Sweden, is given the credit of having instituted what is so well known as the "Swedish Movement Cure;" and in 1813 the Royal Central Institution was established at Stockholm in order that he might practise and teach his system of gymnastics, which were adapted to the well and the sick. Some regarded him as the inventor of this systum of treating certain maladies, while others considered that he only made rational that which had been in use for many centuries amongst the Chinese and other eastern nations. The latter is doubtless the more correct view, for one of his disciples states that Ling thought not, like his predecessors, of merely imitating the gymnastic treatment of the ancients, but he aimed at its reformation and improvement. But the former

view served a useful purpose in stirring up the critics and op-
ponents of Ling's method, who adduced testimony to show that the
method of Ling is that of the Brahmins of India ; is that of the
Egyptian priests ; is that of Asclepiades, of Pythagoras, and of
Herodicus ; is that of which Hippocrates, Celsus, Galen, Rufus
of Ephesus and other physicians, Greek and Roman, have preserved fragments for us ; and that all the movements which Ling
has indicated are described in an ancient book of the Chinese
called the Cong-Fou of the Tao-Ssé.   The critics of Ling doubtless felt roused to righteous indignation, from the following remarks of Mr. Georgii, one of his pupils, who was, perhaps, more
enthusiastic than enlightened : " Let us speak of the series of
movements invented and determined by Ling.   Here the influence comes solely from without, and the patient submits to
the mechanical impression.   Ling means by passive movements
or communicated movements, such as pressures, frictions, per-
cussions, *froissement*, or rumpling of the skin and subcutaneous
cellular tissue, etc., motions and attitudes suitable to produce
temporary or artificial congestion in an organ." (" Estradère
du Massage," 1863.)   However the genius of Ling and the
claims of priority made for him may have been disputed, there
seems to be no doubt as to the merits of the system which he
rescued from oblivion, and to all accounts put upon a scientific
basis.   In the rooms of the Royal Central Institution at Stockholm, persons of every condition and age, the healthy as well
as the sick, executed prescribed movements.   The number
of those who adopted the use of the therapeutic movements increased every year, and among them were even physicians who
in the beginning had been the most opposed to Ling.   In 1844,
the Supreme Medical Board of Russia appointed two members
of the Medical Council to inquire into the merits of the movement and manipulation treatment as practised by M. de Ron,
one of Ling's disciples at St. Petersburg, who had been using it
then for a period of twelve years.   From the highly commenda-
tory report of the councillors we quote the following: "All passive
movements, or those which are executed by an external agent
upon the patient, as well as active ones produced by the effort

of the voluntary muscles, and the different positions with the aid of apparatus or without it, are practised according to a strictly defined method, and conducted rationally, since they are based upon mechanical as well as anatomical principles. *Expe-rience teaches us the usefulness of the institution, as many patients thus treated have recovered their health after having suffered from diseases which could not be cured by other remedies.* We must also mention the testimony of Dr. Bogoslawsky, who himself, after having been cured of a chronic disease, at that institution, has practised diligently this treatment, and who, being appointed consulting physician to that institution, has since then had oppor-tunities enough of observing and witnessing numerous cures." [1]
In most of the large cities of this and other countries, institu-tions similar to the original one at Stockholm are carried on, where movements and stirring up of the external tissues of the body by machinery are successfully employed. In a Japanese book called "San-Tsai-Tou-Hoei," published in the sixteenth century, there is a collection of engravings representing ana-tomical figures and gymnastic exercises; amongst these are pressure, percussion, and vibration, besides passive motion, all of which have been in use with the Japanese from the most re-mote periods. They are employed to dissipate the rigidity of muscles occasioned by fatigue, spasmodic contractions, rheu-matic pains, and after the union of fractures. The Chinese ex-pression *Cong Fou* of *Tao Ssé* is applied to physicians who interest themselves as artists or workmen in mechanical thera-peutics. The Chinese rub the whole body with their hands, press gently the muscles between their fingers, and give a pecu-liar twisting to the articulations. In place of bleeding their patients, they employ kneading and friction to put the blood in motion.
In bringing forward the last evidence to show that Ling's treatment did not originate with him, the history of massage has been allowed to drift from its chronological order. Let us return to it by giving our brief attention to a "quaint and curious volume of forgotten lore" entitled "Illustrations of the

---

[1] Roth on the cure of Chronic Diseases, pp. 18 and 19.

Power of Compression and Percussion in the Cure of Rheumatism, Gout, and Debility of the Extremities, and in Promoting Health and Longevity. By Wm. Balfour, M.D., Author of Illustrations of the Power of Tartar Emetic in the Cure of Fever, Inflammation and Asthma; and in preventing Phthisis and Apoplexy." Second Edition, Edinburgh, 1816. Who can doubt the regularity of Dr. Balfour after reading this? But his brethren did not adopt his views so readily as he thought they ought to have done. The following is a specimen of how he regarded them for their hesitancy: " Medical practitioners encourage their patients in giving perfect rest to parts affected with rheumatism and gout, till, as often happens, they change their actions altogether. It is incumbent on such practitioners to show that there is greater security to life in painful, rigid, and swollen limbs and in frequent and long confinement, than in the free and equable circulation of the blood through every part of the body and in exercise in the open air. It is incumbent on them to show that life is more secure when the functions of the body are imperfectly than when duly performed. It is incumbent on them to show that disease is preferable to health, and more conducive to longevity. . . . It was observed by Lord Bacon that knowledge more quickly springs from absolute ignorance than from error. It is much easier, surely, to instruct the ignorant than to convince the prejudiced. But in spite of the hostility that has been shown to the practice illustrated in the following pages, I have the satisfaction to see it adopted at last by physicians of the first eminence. It may with truth be affirmed that there never was an important improvement in the practice of medicine yet, but what met with opposition. Mercury, bark, cold affusions in fever, vaccination, all experienced it. . . . Percussion and compression have been objected to in gout, on the score of their repelling the disease from the extremities to vital organs. The objection has no foundation whatever, either in matter of fact or in the nature of things. Percussion, instead of repelling, creates an afflux of nervous energy and sanguineous fluid to the part. Vessels in a state of atony are thereby roused to action and circulation is promoted; and bandages support the vessels and enable them to perform their functions. Where fever is present I treat it on general principles." Dr. Balfour claimed for himself the originality accredited to Professor Grosvenor by his friends — that of discovering a new method of treatment — without inquiring if there were any previous data to start from. This fact, however, makes their testimony all the more valuable and unbiassed, which we can doubtless trust, for they were eminent practitioners in their day ; and any one who reads their books would certainly say honest as well. Dr. Balfour's book is mainly made up of reports of cases of rheumatism, gout, neuralgia, sprains, and the results of injuries treated by means of percussion, deep rubbing, and firm compression with bandages. Many of these from a state of chronic suffering speedily got well, and few there were but received some benefit. The cases are well reported and interspersed with forcible and philosophical remarks. Those who claimed originality for using tight strapping in sprains a few years ago would have done well to have first consulted old Balfour.

The *Gazette des Hôpitaux* for 1839 makes known to us that in the island of Tonga, Oceanica, when a person is fatigued from walking or other exercise, he lies down and some of the natives practise divers operations upon him, known under the name of *Toogi-Toogi, Mili* or *Fota*. The first of these words expresses the action of striking constantly and softly with the fist ; the second that of rubbing with the palm of the hand ; the third that of pressing and squeezing the tissues between the fingers and the thumb. These operations are ordinarily done by females ; and they contribute to diminish fatigue and pain, besides producing an agreeable effect which disposes to sleep. When they practise them with the intention of diminishing fatigue alone, the arms and legs are worked upon ; but when there is pain in some place, it is the part affected or the surrounding parts where the operations are applied. In headache the skin over the frontal region and also that over the cranium is submitted to *Fota*, and often with success. Sometimes in cases of fatigue they make use of a process which differs from the proceeding ordinarily employed : three of four little children tread under their feet the whole body

of the patient. The Turks, Egyptians, and Africans, according
to Ardouin, use similar procedures; they rub and press with the
fingers, and they knead all parts of the body. With the Russians,
flagellation and friction by means of a bundle of birch twigs are
resorted to after the subject has been well parboiled in a vapor-
bath. A pailful of cold water is then dashed over him from
head to foot, the effect of which is described as electrifying.
After this he plunges into the snow, and thus tempers himself
like steel to endure with impunity the rigorous climate. The
Siberians and Laplanders also indulge in these luxuries.

It is somewhat remarkable that in France, the country which
first gave massage its name and greatest impulse, this method
of treatment should have become so much neglected. For at
least sixty years the word massage has found a place in the
medical literature of France, but for the past twenty years
very little attention has been paid to it until quite recently.
In the summer of 1884 Prof. Charcot told me that the physicians of Paris did not interest themselves much in massage, but
he hoped that they would. This hope is being fulfilled, thanks
to the labors of MM. Lucas Championnière, Tripier, Rafin,
Norström and others, and the former prestige of the French in
this matter will doubtless soon be regained.

A paragraph from Estradère indicates sufficiently the state
of massage in France in 1863: "Although numerous observations upon the benefits of massage in certain affections have
been communicated to the Academy of Sciences and other
learned societies; although some physicians became alarmed
at the enormous practice of an empiric by the name of Moltenot
who masséed at Orleans in 1833, and entreated the Court of
Justice for a sentence against him; although Recamier and his
pupils, Séguin, Maisonneuve had lectured upon massage before
all the learned societies; although in these times the most
distinguished physicians of Paris very often prescribe massage;
yet for all that it is still under the domain of empiricism,
because physicians are content with indicating its therapeutical
results without interrogating anatomy and physiology for the
reason of these results. Nevertheless this age has a tendency

toward improvement, and already physiologists have given some satisfactory explanations of the effects of massage, passive and mixed movements." With the waning interest of the French physicians in the subject of massage, the Germans and Scandinavians took it up with renewed zeal, and from time to time have furnished instructive accounts of their experiments, successes, and failures. About fifteen years ago, Dr. Mezger, of Amsterdam, treated the (then) Danish crownprince successfully for a chronic joint trouble by means of massage, which he used in a manner somewhat peculiar to himself and in accordance with the teachings of physiology and pathological anatomy. When the prince got well, he sent a young physician to Amsterdam to study Dr. Mezger's method of applying massage, and soon after many old as well as young physicians visited the clinic of Mezger and they all agreed that the so-called massage used in Mezger's manner and according to the indications which a very large experience enabled him to point out, is a most worthy agent in various affections of the joints, besides in inflammations and neuroses. They considered that credit was due to Mezger for having improved massage in a physiological manner, and for having brought it to be acknowledged as a highly valuable method. The esteem in which this method of treatment is held by physicians and surgeons on the continent of Europe who interest themselves in the matter is tolerably well indicated by the following statement from *Schmidt's Jahrbücher :* "It is but recently that massage has gained an extensive scientific consideration, since it has passed out of the hands of rough and ignorant empirics into those of educated physicians ; and upon the result of recent scientific investigations it has been cultivated into an improved therapeutical system, and has won for itself in its entirety the merit of having become a special branch of the art of medicine."[1]

In 1870, Dr. N. B. Emerson gave a very interesting account of the *lomi-lomi* of the Sandwich Islanders. He describes it as

---

[1] Since I first translated the above and put it in print, it has been used to adorn the circular of every humbug in the United States who wishes to make people believe that Turkish-bath rubbing and pounding constitute massage.

a luxurious and healthful form of passive motion which the Hawaiians bestow upon each other as an act of kindness and their crowning act of generous hospitality to a well-behaved stranger. When foot-sore and weary in every muscle so that no position affords rest, and sleep cannot be obtained, it relieves the stiffness, lameness, and soreness and soothes to sleep, so that unpleasant effects of excessive exercise are not felt next day; but in their stead a suppleness of muscle and an ease of joint entirely unwonted. Moreover, the *lomi-lomi* is capable of appeasing and satisfying that muscular sense of *ennui* which results from a craving for active physical exercise. The Hawaiians have an appreciation of the physiological wants of the wearied system which Dr. Emerson thinks it would be well for the people of other civilized nations to imitate. They have various ways of administering the *lomi-lomi*. When one is about to receive it he lies down upon a mat; and he is immediately taken in hand by the *artist* (as Dr. Emerson calls the person who *lomi-lomies*), generally an elderly and experienced man or woman. The process is spoken of as being neither that of kneading, squeezing, nor rubbing, but now like one, and now like the other. Those skilled in the art come to acquire a kind of tact that enables them to graduate the touch and force to the wants of different cases. The natives are such firm believers in it that they sometimes defeat the ends of the surgeon, who would secure perfect rest for fractures, by untimely manipulations. The Hawaiians are a famous race of swimmers and to a foreigner they seem amphibious. When wrecked they sometimes swim long distances, and if one of their number becomes exhausted they sustain him in the water and *lomi-lomi* him at the same time. When he is refreshed by this they all proceed on their watery way together. The people of the Sandwich Islands are of normal stature, strength, and size; but the chiefs are so much larger, handsomer, and more magnificent in muscular development that foreigners would think they belonged to a superior, conquering race did they not know otherwise. The chiefs are about twenty-five per cent. larger than the subjects. The only way in which Dr. Emerson can account for this is

that they are better and more abundantly fed, and have them-
lomi-lomied. How much of the virtues of the *lomi-lomi* are due
to the principles of animal magnetism, Dr. Emerson leaves to
those to determine who are versed in the matter. Who are
they?

Nordhoff, in his book on "Northern California, Oregon and
the Sandwich Islands," published in 1874, gives the following
graphic description of *lomi-lomi:* "Wherever you stop for
lunch or for the night, if there are native people near, you will
be greatly refreshed by the application of *lomi-lomi.* Almost
everywhere you will find some one skilled in this peculiar and,
to tired muscles, delightful and refreshing treatment. To be
lomi-lomied you lie down upon a mat or undress for the night.
The less clothing you have on, the more perfectly the operation
can be performed. To you thereupon comes a stout native with
soft, fleshy hands, but a strong grip, and beginning with your
head and working down slowly over the whole body, seizes and
squeezes with a quite peculiar art every tired muscle, working
and kneading with indefatigable patience, until in half an hour,
whereas you were weary and worn out, you find yourself fresh,
all soreness and weariness absolutely and entirely gone, and mind
and body soothed to a healthful and refreshing sleep. The *lomi-
lomi* is used not only by the natives, but among almost all the
foreign residents; and not merely to procure relief from weariness consequent on over-exertion, but to cure headaches, to
relieve the aching of neuralgic and rheumatic pains, and by the
luxurious as one of the pleasures of life. I have known it to
relieve violent headache in a very short time. The chiefs keep
skilful *lomi-lomi* men and women in their retinues, and the late
king, who was for some years too stout to take exercise, and yet
was a gross feeder, had himself *lomi-lomied* after every meal as
a means of helping his digestion. It is a device for relieving
pain and weariness which seems to have no injurious reaction and
no draw-back but one — it is said to fatten the subjects of it."

There is no longer any doubt about this draw-back which has
been turned into a pull-forward by the eminent neurologist, Dr.
S. Weir Mitchell, of Philadelphia, who in 1877 gave the pro-

fession and the public a careful and interesting account of his successful methods of treating thin, nervous, anæmic and bed-ridden patients, usually women. The methods comprise an original combination of previous well-known agencies: namely, seclusion, rest, and excessive feeding made available by rapid nutritive changes caused by the systematic use of massage and electricity. When "Fat and Blood and How to Make Them" first appeared, not a few men suffering from lack of occupation, and consequent slight or imaginary ills, regretted that they could not have some intractable uterine malady so that they might become interesting and undergo the fashionable and luxurious treatment, which, even without the electricity and imperfectly carried on, has proved an effectual remedy from time immemorial for the depressing influence of the sloth and immorality of the chiefs of the Sandwich Islands. The grandees of Spain are said to be diminutive and decrepit. They do not have any such proceedure as *lomi-lomi* applied to them.

It has always seemed to me that *Nerve and Muscle and How to Strengthen Them* would have been a much better and less sensational title for Dr. Mitchell's book than *Fat and Blood and How to Make Them*, for the reason that there are cases that lose adipose tissue to their advantage under massage. The favorable results of Dr. Mitchell, in the class of cases referred to, have been confirmed by Prof. W. S. Playfair, of King's College, London, and published in a little book under the title of "The Systematic Treatment of Nerve Prostration and Hysteria," 1883. Prof. Playfair clearly shows that certain hitherto intractable cases, usually complicated with primary or secondary uterine trouble, get well or improve remarkably under these general measures of rest and feeding, massage, and electricity administered while the patient is removed from the over-indulgent sympathy and interference of relations and friends. Some of Prof. Playfair's brethren, whose opinion he values highly, think it would be better that patients should remain invalids rather than be cured by such means, referring especially to massage, which to their minds savors too much of quackery. "To my mind," says Prof. Playfair, "quackery does not consist in the thing that is done so much as the spirit in which it is done. The most time-honored and orthodox remedies may be employed in such a manner, and by men boasting of the highest qualifications, as to be fairly chargeable with this taint. That we should be debarred from the use of such potent therapeutic agents as massage or systematic muscular exercise, or electricity or hydro-therapeutics, and the like, because in unworthy hands they have been abused, seems to me almost worse than absurdity." On the other hand, not a few of the profession applaud Dr. Mitchell and Prof. Playfair in a manner that would lead us to infer that they are to be considered the original inventors of massage and give their whole, sole, and exclusive attention to it and the cases in which it is useful. With equal propriety might they be regarded as experts in chemistry and materia medica because they use medicine as well as massage; or specialists in electricity because they have called largely on this medical agent to assist them. Even Dr. Mitchell's own valuable testimony in treating some of the consequences of nerve injuries by massage during the war in the United States from 1860 to 1864 is entirely lost sight of in the burst of blind enthusiasm over his more recent experience. But two points of ebullition need be referred to, for the rest can be judged of from what has previously been said. (1) In the *American Journal of the Medical Sciences*, for January, 1878, the reviewer of *Fat and Blood*, in language more elegant than useful, says: "Although it has been noted by Trousseau that the increased warmth of the skin produced by massage is due to the more active cutaneous circulation, it was reserved to Dr. Mitchell to put this point on an exact scientific basis by a series of accurate thermometric observations." No doubt of the accuracy of the thermometric observations; but what of their value? Let the author speak for himself: "It is well to add," says he, "that the success of the treatment is not indicated in any constant way by the thermal changes, which are neither so steady nor so remarkable as those caused by electricity." However interesting these changes may be, it is worth while to know that they may be safely disregarded in the cases referred to. But this hardly meets our expectation aroused by the statement of Dr. Mitchell that he has some facts to relate in regard to massage which are not known, he thinks, on either side of the Atlantic. (Chap. V. Fat and Food).

(2) Dr. Coghill, in an address before the British Medical Association, said: "It seems to me that the systematic treatment of neurasthenic disorders, practiced with such success by Dr. Weir Mitchell, of Philadelphia, and so recently brought to the notice of the profession with such a corroborative record of success by Prof. Playfair, offers an alternative to Battey's operation of the most promising kind." Prof. Playfair adds: "Here is just that sort of misapprehension which is certain to lead to disappointment, for a purely neurasthenic case is not one for Battey's operation."

General massage for its tonic and sedative effects is almost unknown on the continent of Europe, except in the most ordinary form of rubbing. In the summer of 1889 I could find no one in Amsterdam to give me general massage to rest me from the fatigue of travelling. It has been used more or less skillfully in the city of Boston for the past thirty years.

From this outline of the history of massage we may conclude that, like many other matters in and out of medicine, it has not been steadily progressive, at times being highly esteemed, at others treated with indifference or even contempt, until the weight of eminent authority or the pressure of popular opinion has again raised it from oblivion.

CHAPTER III.

"Vor den Wissenden sich stellen
Sicher ist's in allen Fällen."

A CAREFUL study of the structure of the human body, its con-
tours and conformations, together with the most agreeable and
efficacious manner of applying massage to it, results in proving,
either that the Creator made the body to be manipulated, or else
that He put it into the heart of man to devise massage as a
means of arousing under-action of nerve, muscle, and circulation.
Few there are who have taken any special interest in massage,
but think they have improved it in some way peculiar to themselves, apparently unmindful of the words of the Father of
Medicine, who says that "Medicine hath of old both a principle
and a discovered track, whereby in a long time many and fine
discoveries have been discovered, and the rest will be discovered,
if any one who is both competent and knows what hath been
discovered, start from these data on the search. But whoever,
rejecting these and despising all, shall undertake to search by a
different track and in a different manner, and shall say that he
hath discovered something, will be deceived himself and will de-
ceive others." According to Hippocrates, then, not a few have
deceived themselves and others in the use of massage from want
of starting from previous data on the search. But so long as the
patients are benefited, no harm has come of this. Dr. Mitchell
refers his first interest in this subject to the remarkable results
obtained from its use by a charlatan in a case of progressive paralysis.[1] The description he has given of its mode of application

---

[1] Fat and Blood, p. 51.

35

in "Injuries of Nerves," and in "Fat and Blood," is excellent
so far as it goes, but it is not by any means sufficient.[1]
Dr. Playfair says he never troubles himself as to how massage is
done, and he thinks the details are not of much consequence, provided the operator produce in his patient the waste of tissue
which is essential.   Let Dr. Playfair or any one else become the
patient, and he will be very apt to think in a short time that
the details are of considerable importance so long as the power
of sensation is intact.   If in a sufficient time the results are not
favorable, Dr. Playfair considers the manipulator at fault and
gets another.   When improvement does not follow in due time,
Dr. Otto Bunge, of Berlin, takes a wider view and considers that
either the massage is not properly done, or else that the case is
not a suitable one for such treatment.   He candidly confesses
that he treated his patients too vigorously with massage when
he first tried it, and they left him.

Prof. Playfair thinks that work of this kind ought not to be
expected of the nurse,[2] as she has enough to do in attending to
her other duties; and, moreover, that it requires very intelligent
persons to do massage properly, and even amongst these the
aptitude for the work he finds to be very far from common.
Hence, we may infer that Prof. Playfair considers it of no consequence and also of great consequence as to how and by whom
massage is done.

One author says that massage is difficult to describe, and that
the word embraces too little; another agrees with the first as to
the difficulty, but thinks the word embraces too much; while
neither troubles himself to give an exact outline of the procedure.   To attempt a description of the mode of applying massage
is not an easy matter, and one less experienced in expressing himself than Prof. Playfair might, like him, shrink from the task.
But the writer of this has already made the attempt to describe
general massage in an article published in the *Popular Science
Monthly* for Oct., 1882, and he feels amply repaid for his trouble
in having received the thanks of Prof. Playfair himself.   It is

---

[1] Dr. Mitchell states over his own signature that he does not teach massage.
[2] Applying Lister dressings and many other things spoil the hands for massage.

to be regretted that physicians do not oftener try their hands at massage themselves. They would be fully indemnified for their time and trouble in the improvement of their *tactus eruditus*, which would enable them to appreciate the changes in the tissues brought about by massage, and this would open a new and interesting field of observation to them. Furthermore, the benefit of their visit would then be immediate in place of mediate, as when it is the medicine prescribed and not the physician that does the work ; and a still greater reason is that they would often prevent their glory from departing to another, and that other frequently an ignorant and obnoxious layman, whom the physician is obliged to tolerate or lose the family practice. Visits for massage are not more arduous than many of the visits in surgical, obstetrical, and gynecological practice ; indeed, often less so, besides being much less disagreeable. Physicians daily render service that no menial could be hired to perform. French, German, and Scandinavian physicians often apply massage themselves without any thought of compromising their dignity ; and when such men as Drs. Brown-Séquard, Weir Mitchell, Edward H. Clark, S. G. Webber, and others have sometimes tried their hands at it, I do not see why American and English physicians should not make use of it oftener than they do.

Except among very few, epicures in this matter, if one may say so, there is as yet but little evidence of a desire to place massage, and those who do it, on their merits alone, irrespective of the policy of employing persons who are only rubbing machines, or tolerating peculiar people with strange notions, so long as the poor patients' minds are satisfied. This is too often the case, and then massage is said to have failed and valuable time is lost, whereas, if it had been properly applied, it might have been successful; or, on the other hand, perhaps it should have been omitted altogether, and other remedies employed. In Boston and Philadelphia, and perhaps in other cities as well, efforts have been made by physicians who are thoroughly familiar with massage to instruct intelligent nurses and others how to apply it, and at the training-schools for nurses the pupils receive some general instruction in the matter. In this way, something has been done to bring massage within the rules and regulations of common sense and rational therapeutics. But still there is great room for improvement, even in this direction, for it is but too often the case that, after one or two persons are specially trained to do massage, they are requested to give instruction to some of the pupils at the schools for nurses, and to others; a few of whom after having received some general desultory lessons, are in turn delegated or relegated to teach others, and so on, until by the time massage reaches the needy patients, there is often little left of it but the name. Hence it is not to be wondered at that many a shrewd, superannuated auntie, and others who are out of a job, having learned the meaning of the word massage, immediately have it printed on their cards, and continue their "rubbin'" just as they have always done. An eminent surgeon of Boston advised an equally eminent physician to get a neat Irish girl to rub his leg for a stiffness that followed muscular rupture. The advice was not taken, but enjoyed as a joke. This is but one illustration of many, showing how frequently massage is left to some one who, like Don Quixote's lady Dulcinea del Tobosa, had the best hand for salting pork in all the Tobassas, and charms which, though hid from view, might nevertheless be dwelt on in silent admiration. It is a very common mistake to suppose that those who are of a remarkably healthy, ruddy appearance, plethoric and fat, are the best fitted to do massage. Such people require a great deal of exercise in the open air for the proper oxygenation of their blood, and confining indoor work like massage they soon find to be tedious and irksome. Besides, the stooping attitude and varying positions necessary while doing this sort of work soon put them out of breath; and thus, while suffering from their own ignorance and awkwardness, they fancy they are imparting "magnetism" to their patients at their own expense. Still, I have seen a few stout people who are excellent manipulators. Better that the manipulators should be rather thin, though if of too spare a habit their hands will not be sufficiently strong and muscular, and their tissues will lack that firmness so necessary for prolonged endurance. It has been well said that "those who do massage should be tender and gentle, yet strong and enduring." Those who have a natural tact, talent, and liking for massage, with soft, elastic and strong hands, and physical endurance sufficient to use them, together with abundance of time, patience, and skill acquired by long and intelligent experience, are very useful artists in this department of the healing art; but not likely to be appreciated without the preposterous claims of "magnetism." Dr. E. C. Seguin, in the *Archives of Medicine* for April, 1881, says that even in New York there are few manipulators who can be trusted to do massage well. Non-medical people may become expert and skillful in the individual manœuvres embraced under the term massage, but they ought to have their efforts directed by a physician. Physicians, in addition to want of time, may lack the necessary qualifications for doing massage well; but they would often find it to their advantage to be their own mechanics as well as architects in this, as already intimated.

There are undoubtedly some people who have a natural tact for doing massage, as there are others who have more than usual tact for doing other things, but where fifty or a hundred may claim this tact, there may not be one who really possesses it. It is something that most intelligent people can acquire by instruction and practice, though I have occasionally seen persons who, on account of the natural conformation of their hands, would never make good manipulators. Though Turner, the famous painter, held a professorship of perspective for over thirty years, yet he lectured very little, and then only to the mystification of his hearers, who could make nothing of his blind attempts at explanation. His knowledge of perspective was a matter of intuition and could not be measured by line or rule; and he who could develop perfect distances in his pictures failed to explain them orally. In the practice of massage, the writer knew a character very much like Turner in painting, who, claiming to be the originator of his method, manipulated with a skill surpassing anything he was aware of or could describe. "Entanglement of the nerves," was his diagnosis of every case, and in his opinion he was the only man who could straighten them out. In spite of his having been at variance with all the physicians in Christendom, he had for many years a large patronage; and an account of his successes and failures, if they could be obtained, would be far more interesting and instructive than is Dr. Wharton P. Hood's book "On Bone-Setting, so called," the hero of which explained every joint affection by saying that a bone was out.

The vaguest generalities exist as to the manner of doing massage, even among the best authors on the subject, and, after having studied and tried the methods of all, the writer proposes to formulate, as well as he can, what he has found to be of value, without having adopted the methods of any in particular. But no matter how precisely and carefully worded the description may be, it is not likely to be comprehended unless one sees, feels, and attempts to do massage himself and compares his efforts with others; for massage, though it may be studied as a science, has, like everything else in medicine and surgery, to be practised as an art, and the same may be said of this that Dr. John Hilton said of surgery: there is much that cannot be systematized, that cannot be conveyed from mind to mind in books and articles.

The definition and manner of doing massage is not rendered any clearer by calling slow and gentle stroking in a centripetal direction *effleurage;* or by speaking of deep-rubbing as *massage à friction;* or by using the term *pétrissage* for deep manipulation without friction, or by calling percussion *tapotement.* But custom having sanctioned the use of these words it becomes necessary to mention them.

The multiform subdivisions under which the various procedures of massage have been described can all be grouped under four heads, namely, friction, percussion, pressure, and movement. Malaxation, manipulation, deep-rubbing, kneading, or massage properly so called, is to be considered as a combination of the last two. Each and all of these may be gentle, moderate, or vigorous, according to the requirements of the case and the physical qualities of the operators. Some general remarks here will save repetition: 1. All of the single or combined procedures should be begun moderately, gradually increased in force and frequency to their fullest extent desirable. and should end gradually as begun. 2. The greatest extent of surface of the fingers and hands of the operator consistent with ease and efficacy of movement should be adapted to the surface worked upon, in order that no time be lost by working with the ends of the fingers or one portion of the hands when all the rest might be occupied. 3. If too near the patient, the manipulator will be cramped in his movements ; if too far away, they will be indefinite, superficial, and lacking in energy. 4. The patient should be placed in an easy and comfortable position, with joints midway between flexion and extension, in a well-ventilated room at a temperature from 70° to 75° F. Any sensations of tickling will soon be overcome by the effects of the massage. 5. What constitutes the dose of massage is to be determined by the force and frequency of the manipulations, and the length of time during which they are employed, considered with regard to their effect upon the patient. A good manipulator will accomplish more in fifteen minutes than a poor one will in an hour, as an old mechanic working deliberately will accomplish more than an inexperienced one working furiously. 6. The direction of the procedures should almost invariably be from the extremities to the trunk, from the insertion to the origin of the muscles, in the direction of the returning currents of the circulation. Friction or *effleurage* may be spoken of as circular and rectilinear ; the latter may be vertical or parallel to the long axis of a limb ; or horizontal, transverse, or at right angles to the long axis. Transverse friction is a very ungraceful and awkward procedure. It has been introduced on theoretical considerations alone, and may without loss be laid aside. A slight deviation from the method ordinarily recommended in doing straight-line friction I have found to be more advantageous ; for though in almost every case the upward strokes of the friction should be the stronger, yet the returning or downward movement may with benefit lightly graze the surface, imparting a soothing influence, without being so vigorous as to retard the circulation pushed along by the upward stroke, and thus a saving of time and effort will be gained. The manner in which a carpenter uses a plane represents this forward-and-return motion very well. In giving a general massage, it is immaterial whether the upper or lower extremities be done first. Let us begin with the hands, and here a convenient extent of territory is from the ends of the fingers to the wrist, each stroke being of this length, the returning stroke being light and without removal of the hand. The rapidity of these double strokes may be from ninety to one hundred and eighty per minute. The whole palmar surface of the fingers in easy extension should be employed, and in such a manner that they will fit into the depressions formed by the approximation of the phalanges and metacarpal bones, the patient's hand meanwhile resting in the other hand of the manipulator. The heel of the hand should be used for vigorous friction of the palm, done by a semi-circular pushing movement, and the same can be done to the sole of the foot with somewhat less of the semi-circular motion. The effect of this, when well done, is remarkably agreeable. The right hand of the operator should be used for the right hand and foot of the patient, and the left for the left, for in this manner they fit each other better. From the wrist to the elbow, and from the elbow to the shoulder-joint, are each suitable extents of surface to be worked upon, and here not only straight-line friction, extending from one joint to another, may be used, but also circular friction. The form of the latter, which appears to me most serviceable as it includes the advantages of the other two, is in that of an oval, both hands moving at the same time, the one ascending as the other descends, each stroke reaching from joint to joint, the upward being carefully kept within the limits of chafing the skin, while they move at a rate of from seventy-five to one hundred and eighty each per minute, or one hundred and fifty to three hundred and sixty with both hands. It is well to begin these strokes on the inside of both arms and legs, so that the larger superficial and deep vessels may be first acted upon, as this influence extends at once, though indirectly, to their tributaries and ramifications. But it is not always practicable to place the hand of the patient on a support so that the operator can work with both hands on the arm. But then he can grasp the patient's right hand by its dorsum with his left while his other does oval friction on the anterior aspect. And for the back of the arm the manipulator will grasp the patient's hand as in the act of shaking hands while his disengaged hand does the friction.

Time, effort, and effect will be made the most of by doing friction upon the foot with the hands at right angles to it, one hand upon the dorsal aspect, and the other upon the sole, moving alternately and in a circular manner, the one ascending as the other descends. Around and behind the malleoli will require a special pushing stroke with the fingers, and for this the operator will unconsciously change his position so as to face the patient. As the lower limbs are larger than the upper, the lateral and posterior aspects from ankle to knee will form a convenient territory, while the lateral and anterior aspects will make another for thorough and efficacious friction. This will be best done with the knees semiflexed and the operator standing facing the patient for the posterior and lateral aspects, and after having completed the friction here, without stopping the strokes, he will turn with his back to the patient and continue the stroking on the anterior and lateral aspects, each thumb following the other with tolerably firm pressure over the anterior tibial group of muscles; but owing to the latter position of the operator only upward friction can be done without the light downward stroke. The same systematic division of surface may be made above the knees as below, with the addition of another formed by the inner and anterior aspect of the thigh, and they may be dealt with in like manner; but the operator's back to the patient will on the whole be the easiest and most efficacious way of applying friction to the thighs. The number of strokes below the knees will vary from one hundred to one hundred and sixty with each hand; above from sixty to one hundred. From the base of the skull to the spine of the scapula forms another region naturally well bounded for downward and outward semi-circular friction, and from the spine of the scapula to the base of the sacrum and crest of the ilium forms

another surface over which one hand can sweep, while the other
works towards it from the insertion to the origin of the glutei
at an average rate of sixty or seventy-five a minute with each
hand for a person of medium size. It will be observed that on
the back and thighs the strokes are not so rapid as on the other
parts mentioned, for the reason that the skin is here thicker and
coarser, in consequence of which the hand cannot glide so easily,
and the larger muscles beneath can well bear stronger pressure;
besides, the strokes are somewhat longer, all of which require
an increased expenditure of time. The chest should be done
from the insertion to the origin of the pectoral muscles, and the
abdomen from the right iliac fossa in the direction of the
ascending, transverse, and descending colon. But in these
situations friction is seldom necessary, for the procedure about
to be considered accomplishes all that friction can do and a
great deal more. The force used in doing friction is often much
greater than is necessary, for it should only be intended to act
upon the skin, as there are better ways of influencing the tissues
beneath it. If redness and irritation be looked upon as a
measure of the beneficial effects of friction upon the skin, then
a coarse towel, a hair mitten, or a brush would answer for this
purpose a great deal better than the hand alone. But for intelligent variation of pressure, agreeableness of contact, and adapt-
ability to even and uneven surfaces, no instrument has yet been
devised to supersede the human hand. In union there is
strength, and the fingers should be kept close together in doing
friction and manipulation. But it is astonishing how persistently
they are sometimes held .out straight and spread far apart,
reminding one of the feet of a frightened duck, and the sound
of quack suggests itself as appropriate for the one as the other.
This would be still more appropriate when, as often happens,
the hands are made to traverse the air to an undue extent, accompanied with a snapping of the fingers, reminding us of Mrs.
Boffin's horses that stepped higher than they did long ways;
and if with these ungraceful flourishes perspiration be mistaken for inspiration, and blind enthusiasm for "magnetism,"
there can be no doubt of the genus to which the operator
belongs.

The useless flourishes of many while doing friction might impress an uninitiated spectator as evidences of expertness. They bear the same relation to massage, pure, effectual, and agreeable, that the superfluities of architecture, known as the Queen Anne style, wherein comfort is sacrificed to beauty, bear to the classic detail of Greek architecture.

The advantages of ordinary rubbing are not to be despised, and by many this is supposed to be all there is to massage; but it is the least essential part of it. One of the old French dictionaries says there is reason to believe that massage has upon the skin the advantage of friction, that it acts above all upon the more deeply situated tissues, etc., thus implying that massage, properly so-called, is something different from friction, and yet has the same effect upon the skin, while exerting a more extended range of influence. By this, we understand massage proper to be manipulation, deep-rubbing, kneading or malaxation, which is certainly the most important, agreeable, and efficacious procedure of all. It is done by adapting as much as possible of the fingers and hands to the parts to be thus treated, and, without allowing them to slip on the skin, the tissues beneath are worked upon in a circulatory manner by a sort of kneading, rolling, squeezing, manipulatory motion, proceeding, as in friction, from the insertion towards the origin of the muscles, from the extremities to the trunk. For this purpose the same divisions of surface as for friction will be found most convenient. Beginning then with the fingers from the roots of the nails, the thumb of the manipulator will be placed on one of the fingers of the patient, and parallel to it, while on the opposite surface, the second phalanx of the index finger will be simultaneously placed at right angles to this, and between the two, the finger of the patient will be compressed and malaxated at the rate of from seventy-five to a hundred and fifty per minute. The dorsal and palmar surfaces will of course receive special attention, while the lateral aspects will come in for a secondary share. If the manipulator be sufficiently expert, he can work with both hands on this small surface, or he can take one of the patient's fingers in each of his own hands and proceed with the

same rapidity as with one. Each finger and thumb will be taken
in turn, and the manipulations extended over the metacarpal
and carpal bones as far as the wrist-joint, and finally the palm
of the hand will be done by stretching the tissues vigorously
away from its median line. Each part included in a single grasp
may receive three or four manipulations before proceeding onward
to the adjacent region. The advance upon this should be such
as to allow the finger and thumb to overlap one-half of what has
just been worked upon. Advance and review should thus be
systematically carried on, and this is of general application to all
the other tissues that can be *masséed*. The force used here and
elsewhere must be carefully graduated so as to allow the patient's
tissues to glide freely upon each other, for, if too great, the movement will be frustrated by the compression and perhaps bruising
of the tissues ; if too light, the operator's fingers will slip ; and if
gliding with strong compression be used, the skin will be chafed.
To avoid this last objection, various greasy substances have been
employed so that ignorant, would-be masseurs may rub without
injuring the skin. When the skin is cold and dry, or cold and
moist, and the tissues in general are insufficiently nourished, as
well as in certain fevers and other morbid conditions, there can
be no doubt of the value of inunction ; but no special skill is required to do this, and there is no need of calling it massage, unless it be to please the fancy of the patient. Removal of hair is
entirely unnecessary. Massage can be done as effectually on the
head as on any other part.

The feet may be dealt with in much the same manner as the
hands ; using the ends of the fingers to work longitudinally between the metatarsal as between the metacarpal bones, and the
tissues of the sole should be stretched vigorously away from the
median line ; and lastly the heel, accurately adapted into the palm
of the hand and between the thenar eminences and fingers, will be
worked upon in a squeezing, circulatory manner. Upon the
arms and legs and, indeed, upon all the rest of the body, both
hands can be used to better advantage than where the surfaces
are small. Each group of muscles should be systematically
worked upon, and for this purpose, one hand should be placed

opposite the other, and where the circumference of the limb is
not great, one hand may be placed in advance of the other, the
fingers of one hand partly reaching on to the territory of the
other, so that two groups of muscles may be manipulated at the
same time with grasping, circulatory, spiral manipulations, one
hand contracting as the other relaxes, the greatest extension of
the tissues being upward and laterally, and on the trunk, forearms, and legs away from the median line. It is needlessly wearisome to both patient and manipulator if the hands are kept
closely adapted to a limb its whole length in doing this vermicular squeezing, besides it interferes with the circulation. To
avoid this, it is only necessary to raise the hands slightly in ad-
vancing. Subcutaneous bony surfaces, as those of tibia and
ulna, incidentally get sufficient attention (unless œdema be present) while manipulating their adjacent muscles, for if both be
included in a vigorous grasp, unnecessary discomfort results.
Care should be taken not to place the fingers and thumb of one
hand too near those of the other, for by so doing their movements would be cramped. With the fingers and thumbs at proper distances from each other, not only are the tissues immediately under them acted upon, but those between them are
agreeably stretched. The advance should be upon the previously
unoccupied stretched region. Space and force will be indicated
by the elasticity, or want of it, in the patient's tissues, the object
being to obtain their normal stretch, and in this, every person is
a law to himself, the character of tissues varying with the
amount and quality of adipose, modes of life, habits of exercise,
etc. A frequent error on the part of the manipulator is in attempting to stretch the tissues in opposite directions at the same
time, especially at the flexures of the joints, where the skin is
delicate and sensitive, and where the temptation to such procedure is greatest, because easiest, the effect being a sensation of
tearing of the skin. It is well to go over a surface gently and
superficially before doing the manipulation more thoroughly and
in detail. In the case of the arm, the two hands will embrace
the whole circumference, the thumbs occupying the median
line, on the anterior aspect as well as upon the back of the

arm. The supinators should receive a special malaxation with
the grasp of one hand. Above the elbow, one hand will seize
and squeeze the biceps while the other takes the triceps. The
median portion of the deltoid will receive most thorough attention from the thumbs placed parallel to its fibres, while the palms
and fingers are engaged with the anterior and posterior aspects
of the muscle, and after this its margins and the whole muscle
can be well worked by seizing the muscle with the hand at right
angles to its fibres. In manipulating a leg of considerable size,
three divisions of surface will be found necessary : the posterior
and lateral aspects will form one ; the stretching of the peroneal
muscles from those of the anterior tibial region, which is done
by placing one thumb in advance of the other on the outer side
of the fibula, and alternately rolling the muscles away from each
other, will make another ; and for the third the thumbs will
be placed upon the tibialis anticus and a simultaneous rolling of
the tissues will be made away from the crest of the tibia.
In all of these procedures, no parts of the hands need be idle, for
when not specially occupied, they can be giving secondary attention to the surfaces they cover. Of course, if the limb is small
it can all be *masséed* at once in the grasp of the two hands ; but
even in this case, when special massage is required, these three
divisions are necessary. The cushions of the thumbs, the heel
of the hand, the thenar and hypothenar eminences fit admirably
into the depressions of the joints, especially those of the ankle,
knee, and elbow, while the rest of the hand is occupied with the
adjacent tissues. Above the knee, one hand will grasp the
adductors while the other embraces the quadriceps extensor,
and the alternate contraction and relaxation of the hands will be
made in such a way as to stretch these two groups of muscles
away from the line of the femoral artery. The posterior
femoral region may next be gone over, which will principally engage the fingers, while the upper parts of the hands
work upon the sides of the limb. The external aspect of
the thigh may receive as vigorous kneading as it is possible
to give with evenly distributed force ; and with the thumbs
in advance of each other, on the rectus femoris more special and

effectual manipulation can be given to the extensors, while the remaining surfaces of the hands make a review of the lateral aspects of the thigh. The rate of these manœuvres varies from seventy-five to a hundred and fifty with each hand per minute on the arms, from sixty to ninety on the legs, and from forty to eighty on the thighs, where more force is required on account of the larger size and density of the muscles, and the need of using sufficient force to extend beneath the strong, tense fascia lata.

On the back the direction of these efforts will be from the base of the skull downward, stretching the tissues away from the spinal column while manipulating in graceful curves at an average rate of sixty per minute with each hand. If this be done on one side of the back, as it most frequently has to be, while the patient lies on the other side, it is one of the most difficult manœuvres for beginners to learn, and some never succeed in acquiring it. While both hands are at work on separate spaces occupied by each, the one follows the other, not in an opposite, but in the same circular manner alternately, the one contracting as the other relaxes. And here one hand can often be reenforced by placing the other upon it, and thus massage may be done with all the strength the manipulator can put forth. The position of the shoulder blades is important, for if the upper arm be parallel with the side, then the posterior border of the shoulder blade will be so near the spinal column that scarcely any space will be allowed to work upon the muscles between the scapula and spine. If the upper arm be stretched forward its full length, then the superficial muscles between the spine and the scapula will be so tense that those beneath cannot be effectually reached by masssge. Hence the arm should be placed midway between these two positions. With the ends of the fingers the muscles on each side of the spinal column can be rolled outwards, and the supra-spinous ligament can be effectually *masséed* by transverse to-and-fro movements. The ends of the fingers and part of their palmar surface should also be placed on each side of the spinous processes, and the tissues situated between these and the transverse worked by up-and-down motions parallel to the spine, taking care to avoid the too frequent error of making pushing, jerky movements in place of smooth, uniform motions in each direction.

On the chest and abdomen the same general direction will be observed as in using friction, but the manipulation will be more gentle than on the back and limbs, for the tissues will not tolerate being so vigorously squeezed and pinched. Here the massage will consist of moderate pressure and movement with the palms of the hands, and rolling and grasping the skin and superficial fascia; and, after this, on the abdomen, firm, deep kneading in the direction of the ascending, transvere, and descending colon, using for this purpose the greatest force with the heel of the hand on the side of the abdomen next the operator, and on the other side the strongest manipulation with the fingers, avoiding the frequent and disagreeable mistake of pressing at the same time on the anterior portions of the pelvis. The sides will incidentally receive sufficient attention while the back, chest, and abdomen are being manipulated. When constipation is obstinate, it is a good plan to commence manipulation of the abdomen over the left venter of the ilium and work so as to push the contents of the descending colon towards the rectum; then begin again a little further backwards on the colon and work in the same direction as before, attempting to unload the large intestine, and so on until the whole colon is traversed back to the ileo-cæcal valve, and again from there to the sigmoid flexure of the colon.

Tolerably fair and rapid rolling of the muscles of the back may be done by means of a rubber roller about three inches in length, and one and a quarter inches in diameter, secured to a handle in the manner of a printer's ink-roller. In 1880, a homœopathic physician of New York claimed the honor of having invented an electro-massage instrument in which a roller electrode was made to cause rotation of a pair of helices near a magnet, and thus give rise to an induced current. The current from this large metallic roller, three inches in diameter, was painful and jerky, and such as no mortal patient who knew anything about massage would endure, unless to escape the penalty of death. It certainly was nothing like massage, and as the *Archives of*

Medicine for April, 1881, said, "its utility remained to be determined." In October, 1881, it was advertised that in consequence of the great demand for it the price had been reduced forty per cent., and this sort of demand has doubtless increased so that in all probability it can now be given away.

The wire of either pole of any battery can easily be attached to a conducting roller with a non-conducting handle. If the utility of this should ever be demonstrated, a roller much smaller than three inches in diameter would be more convenient for the operator and agreeable to the patient. The smaller the roller, the more rapidly it would revolve on being pushed, and the greater would be the mechanical impression. Zabludowsky scorns the idea of doing anything worthy of the name of massage with instruments or machinery, and emphatically states that only in the hands of physicians can it prove an effectual curative means. The sponges or poles of any battery may be pressed and moved so as to give a kind of massage while the current is passing. This is much more agreeable and effectual than a current from a large metallic roller.

Before leaving this part of the subject, the writer begs leave to say something more about the common errors into which manipulators fall, even some of those who pass for being skilful. Many do not know how to do the kneading or malaxation with ease and comfort to themselves and to their patients, for in place of working from their wrists and concentrating their energy in the muscles of their hands and forearms, they vigorously fix the muscles of their upper arms and shoulders, thus not only moving their own frame with every manipulation, but also that of their patients, giving to the latter a motion and sensation as if they were at sea in stormy weather. By this display of awkward and unnecessary energy, not only do they soon tire themselves out and fancy that they have lost magnetism by imparting it to their patients, but by the too firm compression of the patient's tissues they are not allowed to glide over each other; and hence such a way of proceeding entirely fails of the object for which it is intended. Surely, cultivation is the economy of effort, and the most perfect art consists in acting so naturally that it does not

appear to be any attempt at art at all. The following words of
J. Milner Fothergill are here applicable: "The knowledge
which one man acquires by the sweat of his brow after years of
patient toil and painstaking cannot be transferred in its entirety
to another. Individual acquired skill cannot be passed from
brain to brain, any more than the juggler who can keep six balls
in the air, can endow an on-looker with like capacity, by merely
showing him how it is done. The muscles, and still more their
representatives in the motor area of the brain hemispheres, require a long training before this manual skill can be acquired."

Friction and manipulation can be used alternately, varied with
rapid pinching of the skin and deeper grasping of the subcutaneous cellular tissue and muscular masses, and when necessary
with percussion, passive, assistive or resistive movements, finish-
ing one convenient surface or limb before passing to another,
and occupying from half an hour to an hour with all or part of
these procedures. Pinching is rather an agreeable way of exciting the circulation and innervation of an inert skin, and for
this purpose it is best done rapidly at the rate of one hundred
to one hundred and twenty-five per minute with each hand.
The grasp of a fold of skin should not be relaxed until seized by
the finger and thumb of the other hand. To act upon the subcutaneous cellular tissue, a handful of skin is grasped and rolled
and stretched more slowly than by the preceding method. A
deeper, momentary grasping of the muscles is often advantageous,
and may be called a *mobile intermittent compression*, and this,
indeed, is what the whole of massage strictly speaking consists
of. Percussion, in general only applicable over muscular masses,
may be done in various ways. In the relative order of their
importance they are as follows: 1. With the ulnar borders of
the hands and fingers. 2. The same as the first, but with the
fingers separated so that their adjacent sides will strike against
each other like a row of ivory balls. 3. With the ends of the
fingers, the tips being united on the same plane. 4. With the
dorsum of the upper halves of the fingers loosely flexed. 5.
With the palms of the hands. 6. With the ulnar borders of the
hands lightly shut. 7. With the palms of the hands held in a

concave manner, so as to compress the air while percussing. The back of a brush or the sole of a slipper sometimes answers very well for percussion; but still better are India-rubber air balls secured to steel or whale-bone handles. With these, one gets the spring of the handles together with the rebound of the balls, and thus rapidity of motion with easily varying intensity may be gained, if the operator knows how to let his wrists play freely, as he should do in all the different ways of percussing. The number of blows may vary from two hundred and fifty to six hundred with both hands. The blows should be smart, quick and springy, not solid and hard, and they should be transversely to the course of the muscles with the ulnar border of the hand and palmar surface ; except in the case of the back, which may not only be percussed with the hands at right angles to it while the patient is lying, but still more effectually when the patient is standing bent forwards, so as to put the dorsal muscles on the stretch. The operator's hands are then most easily parallel to the spine, and can rapidly strike the muscles on each side of it, causing, we have reason to suppose, a vibratory effect, as when the string of a bow is vibrated. Moreover, in this position, the muscles, being tense, protect the transverse processes from the impact of the blows which is communicated to the nerves as they emerge from the intervertebral foramina, and the effect is usually perceived to their distribution as a peculiar and delightful thrill. Percussion must be carefully used, or it will leave the muscles lame and sore.

Remedial movements have been more fully than clearly described in books on "Movement Cure." A comparison of different ways of executing them demonstrates that the part of the limb or body taken hold of for leverage, and the manner of seizing the same, the direction of resistance and force opposed are all of importance in order that the movements may be done easily, efficaciously, and harmoniously. Those who would apply them should know the anatomy and physiology of the joints and their natural limits of motion. Except in the case of relaxed joints, passive motion should be pushed until there is a feeling of slight resistance to both patient and manipulator ; for by this it will be known that in healthy joints the ligaments, capsules, and attachments of the muscles and fasciæ are being acted upon. Resistive movements are such as the patient can make while the operator resists; or such as the operator over-comes when the patient resists, as when a group of muscles is voluntarily contracted, the operator extends them. The former have been called double concentric movements, and the latter double eccentric. It seems to me that the author of these terms must have been somewhat eccentric, for even such a good writer as Estradère pardonably confounds their meaning, as can be seen by comparing his explanation of them on page 72 with that on page 80 of his book on "Massage." Brown-Séquard first pointed out the fact to me that when it is desirable to exercise a group of very much enfeebled muscles, if they be first con-tracted to their utmost, it will require much greater force to overcome this contraction, than they could overcome in passing from a state of relaxation to contraction, and I have since proved the practical value of this suggestion. Most frequently, however, it will be necessary to offer resistance against the patient's movements, and then the opposing force should be carefully and instinctively kept within the limits of the patient's strength, so that he may not recognize any weakness ; and this, with all these other manœuvres, should stop short of fatigue, at least fatigue that is not soon recovered from. To alternately resist flexion and extension is the *pons asinorum* of manipula-tors, and in a considerable experience of teaching massage I have found but few who could learn to do it well, and many who could not learn to do it at all. Many a patient who has re-covered from an old injury is still as much incapacitated as ever, from the fact that his latent energies can only be dis-covered and made available in this manner. Midway between passive and resistive movements, in the course of certain re-coveries, stand assistive movements. They are but little under-stood and seldom used. Let it be supposed that, in the absence of adhesions and irreparable injury of the nerve centres, the deltoid has but half the requisite strength to raise the arm. So far as any use is concerned, this is the same as if there were no power of contraction left in the muscle. But, if only the other half of the impaired vigor be supplemented by the carefully graduated assistance of the operator, the required movement will take place; and in some cases, if this be regularly persisted in, together with manipulation and percussion, more vigorous contraction will be gained, and, by-and-by, the patient will exert three-fourths of the necessary strength, and later the whole movement will be done without aid; and, as strength increases, resistance can be opposed to the movement. The importance of these measures can hardly be over-estimated in cultivating the strength of weakened muscles, while at the same time finding out how much they can be used. Still another kind of movement may be spoken of — namely, vigorous passive motion — with a view to breaking up adhesions in and about joints. It is the secret of success and of failure of the people who call themselves "bone-setters," the methods of whom have been well studied and explained by Dr. Wharton P. Hood, of London, in his highly entertaining book, "On Bone Setting, So-called."

So much for a general outline of movements. Let us speak of them more in detail. In doing a resistive movement in which the patient is the prime mover, the operator waits till he finds the movement begun, then gradually increases the resistance to the utmost within the limits of the patient's strength, and finally slacks up more slowly. This must be practised by the operator on well people until he can instinctively judge of the patient's strength and make elastic resistance. The resistance must be in line with the patient's movements, and the grasp of the operator must not be so firm as to interfere with his own sensation or that of the patient. It will often be found that the patient uses nearly all his strength in contracting his muscles, and scarcely any in overcoming the resistance, in which event it will be necessary to tell him to move more quickly and not try so hard. Here physiology steps in and gives us a reason for the faith that is in us, showing how science agrees with art. Muscular contraction presents three phases: 1. A preparatory or latent period during which there is no visible movement when the nerve and muscle are getting ready to act. 2. A phase of shortening or contraction. 3. That of relaxation or return to its former length. In harmony with these phenomena, and with the manner of doing each and all of the manipulations, and especially resistive movements, physiology teaches us that at the close of the latent period the muscle shortens in each fibre, at first slowly, then more rapidly, and lastly more slowly again. In accordance with these physiological principles of muscular contraction, it would be difficult to conceive of anything that would make graduated and harmonious resistance, save human power guided by human intelligence. Springs and elastic contrivances come nearest to it, and do very well on starting; but the longer the pull or push, the stronger becomes the opposition, and there is no third stage of lessened resistance.

The manner of taking the hand to give it passive motion of flexion and extension, and to resist flexion, is the same. Let the patient's forearm be midway between pronation and supination, and then seize the hand as if about to shake hands, the right hand for the right hand of patient, or the left for the left, so as to bring the resistance on a line with the metacarpo-phalangeal joints, which affords the best leverage for both patient and operator; the other hand at the same time will support and make counter-resistance on the back of the arm about one inch above the wrist. To resist extension of the hand, the patient's forearm should be pronated, then the operator will take the hand in such a way as to bring the resistance over the heads of the metacarpal bones, his right hand for the patient's left, and the left for the right, while the other supports and steadies the arm above the wrist on the anterior surface. For passive pronation and resistive supination the manner of holding the arm is the same ; the operator's right hand seizes the left wrist and lower ends of the radius and ulna of the patient so that the metacarpophalangeal joint of his thumb is upon and behind the styloid process of the radius, the point of resistance, care being taken not to squeeze so tightly as to prevent these bones from rotating upon each other ; in the mean time the other hand of the operator gently supports the arm of the patient. For passive supina-

tion or resistive pronation the same grasp suffices, with the right
hand of the manipulator for the right arm of the patient, or the
left for the left, which seizes the wrist and lower ends of the
radius and ulna so that the metacarpo-phalangeal joint of the
thumb is anterior to the styloid process of the radius, the same
care being observed not to hinder the motion by holding too
tightly, while the arm of the patient rests in the other hand of
the operator. The right wrist of the patient is gently held by
the right hand of the operator, while the left hand steadies the
arm just above the condyles of the humerus in doing passive or
resistive motion of the forearm, or the passive combined motion
of flexion, extension, pronation and supination, abduction and
adduction, together with rotation of the humerus, all of
these seven last movements being accomplished at one and the
same time by simply making the wrist describe a circle. Circumduction of the humerus is most easily and effectually done
by standing behind the patient, and while fixing the right shoulder with the left hand, or the left with the right, the other hand
takes the arm just below the elbow and makes this traverse as
great a circle as moderate resistance will allow, the operator
remembering that the greatest resistance will be at the upper
and outer third of the circle, owing to the natural formation of
the joint. The same hold and support answer well for resisting
a forward motion of the upper arm. If the patient be lying on
the right side, or the operator be standing in front of the patient
while the latter is sitting, tolerably good circumduction may be
done by taking the left wrist in the left hand and placing the
right hand upon the elbow. But this is not so effectual as the
first method, owing to the great mobility of the scapula. Backward motion of the humerus can be steadily and definitely resisted by taking the right hand of the patient in the right of
the operator, or the left with the left, while the other is placed
above and behind the elbow. The action of the deltoid in elevating the arm can be well resisted by steadying the shoulder
with one hand while the other is placed on the outside of the
upper arm, and the opposition can easily be increased by moving
the hand towards the elbow, or diminished by moving the hand

towards the shoulder, the operator meanwhile standing behind the patient. When it is desired to limit motion to one joint, it will be observed that the proximal side should be steadied while the distal side is moved and nowhere is this more disregarded than with the fingers.

For passive or resistive motion of the ankle-joint, the best way of taking hold is not by seizing the heel with one hand while the other surmounts the toes, as is generally done, but with the right hand for the right foot, or the left hand for the left foot, by grasping the metatarso-phalangeal joints at right angles while the other hand supports the leg above the ankle. For this purpose the operator should sit facing the patient and be careful that his active arm is in a straight line with the patient's movement. This affords the best leverage for flexion and extension of the foot, as well as for a circumductory motion by making the place of seizure describe a circle, the outer half of which will offer the greatest resistance, owing to the large internal lateral ligament and the stronger structures on the inside of the joint. The same hold answers for resisting flexion and extension of the foot. When this is done alternately, in the interval of change, here and elsewhere, the hand of the operator must alter its position slightly so as to present a proper surface for resistance. In the case of the foot and forearm, the fingers will pull and resist flexion, and the heel of the hand will push against extension. On the foot the tendency is to make resistance too near the toes; opposite the heads of the metatarsal bones on the back and sole are the points that afford the best and most natural leverage. By seizing the heel and holding the ball of the foot as just described, for passive motion, a twisting motion can be given to the whole foot which acts more decidedly on the tarsal and metatarsal articulations. Flexion and extension of the leg at the knee, either passively or resistively, are seldom necessary to be done alone (except for some special reason), as they are accomplished so much better together with flexion and extension of the thigh; and for this purpose the right heel of the patient is taken in the palm of the right hand of the operator, or the left in the left, while the other hand holds the calf, and a steady, uniform push is made, the limb, by its own resiliency, usually returning to a state of extension. Circumduction of the thigh will be performed by simply changing the hand that holds the calf on to the top of the knee, which affords excellent and easy leverage. On each side of the forefinger of the hand that manages the heel, the covering will be held by the thumb and middle finger so that the patient will not be fanned into the next world. Opposing flexion and extension of the leg and thigh may be done by holding the leg and foot as for passive or resistive movements of the ankle, and if the couch on which the patient lies be of ordinary height, the operator will often require to rest on the knee next the patient, and in resisting extension will throw the weight of his body in part or wholly against the extending limb; and in doing this the arm must not be extended but flexed, so as to bring the hand as near as possible to the shoulder in order that the resistance may be strong and steady by having the weight of the body added to it; or he can stand with his back to the patient and clasp his hands on the sole of the arch to resist extension. Opposing abduction and adduction of the thighs scarcely needs mention, so simply is it done by alternately placing the hands on the outer and inner aspects of the semi-flexed knees; and to resist the contraction of the psoas magnus and iliacus internus alone, resistance may be made to the flexing thigh on any part of its anterior aspect.

Passive stretching of the arms and shoulders, of the pectoral muscles and latissimus dorsi, can be done agreeably and effectually while the patient lies squarely on the back, the head and shoulders being slightly elevated on an inclined plane. The arms of the patient are extended upwards on a line with the body, and the manipulator standing behind, holding the hands, makes a gentle, elastic, and vigorous pull; and if the feet be held, a stretch of the trunk and lower limbs can also be obtained.

The manner of seizing the hands of the patient for this purpose is worthy of particular notice. They are grasped so that their palmar surfaces obliquely cross the palmar surfaces of the operator, the fingers of the operator surround the metacarpal

region of the thumb, while the thumb of the operator passes
between the thumb and index finger of the patient, and the heel
of the hand rests securely upon the metacarpal region of the
patient's little finger, so that the hands of the patient and manipulator are complementary to each other. This is a puzzle for
most people to do, even after having seen it done. The same
hold suffices for resisting a downward pull of the arms which
brings the aforesaid muscles more strongly into play, elevates
the chest and deepens inspiration. With the patient sitting
slightly inclined forward, the hands clasped at the back of the
head, the oblique and transverse muscles of the abdomen can
be passively exercised by seizing the patient at or near the
shoulder-joints and rotating the body, the operator, of course,
standing behind the patient. The same position of the patient
does well to make these muscles act more vigorously by opposing their voluntary contraction. In doing this the operator
stands behind and to one side of the patient, steadying the body
of the latter with the left hand upon the left shoulder, or the
right upon the right, at the same time that the other hand holds
the humerus near the elbow, by which great leverage is obtained
in resisting rotation of the trunk. At first the patient will
naturally err in limiting the motion to the arms and chest, but
he can be gradually educated to lessen this and increase the
rotation at the waist. Upon a vigorous and healthy tone of the
muscles of the abdomen depends to a large extent the welfare
of the organs situated beneath them, and no muscles are so
much " left out in the cold " for want of exercise as these.
Gentle rowing exercise for the muscles of the back can be given
to invalids by standing in front of them and taking hold of the
hands ; but for this purpose elastic tubes or straps answer well,
as the weight of the body makes the pull strongest at its termination. Other movements, passive and resistive, may be devised to meet the indication of individual cases, and of course it
will not be forgotten that active or voluntary movements may
be turned to good account — with special modifications — as
remedial agents.

Gymnastics and calisthenics do not belong to a description

of massage. In accordance with the principles here laid down for doing massage and resistive movements, we find but one plate in Schreiber on Massage at page 5, and that is borrowed from the accomplished oculist Pagenstecher, of Wiesbaden. All the other plates show more or less disregard for economizing time, space, effort, and the comfort of the patient, the most remarkable being on page 48, where the patient has to hold on to the chair upon which he sits and fixes all the muscles of both arms, chest, and trunk for the simple purpose of flexing his right arm against resistance. If he were driving a runaway horse and in danger of being thrown from his seat, he could scarcely be better represented. In Laisné on Massage, there is but one plate, at page 110, that at all agrees with the mode of proceeding I have described, and in this but one-half as much of the hand is occupied as might be ; while his most striking disregard for common sense appears in plate 5 on page 48, where one hand is placed in the popliteal space, thus hindering the very movement the other is trying to make — namely, flexion of the leg. Between these two extremes represented by Schreiber, Laisné, and others, all sorts of anomalies occur, so that we may fairly conclude that massage and remedial movements are usually done "every which and t'other way" in hopes of hitting upon a right way ; but from what has been said in this chapter, we have a right to expect and require a tolerable degree of precision. Cures by massage are not sinecures, but the task of applying it is rendered much greater than is necessary from the crude notions that prevail on the subject; and even Dr. Weiss, of Vienna, recommends the physician to drink a glass of good old wine every fifteen minutes if he himself undertakes such arduous work.[1] This, if furnished at the patient's expense, might sometimes be an inducement that would prolong the treatment unnecessarily. But such suggestions proceed from the same misconceptions of massage as require the removal of hair and the constant use of grease, and those who cannot do massage without them certainly cannot

---

[1] Wiener Klinik, page 335, 1879.

do anything worthy of the name with them. People of education, refinement, and delicacy of touch can do massage a great deal better than those who rely upon the unguided strength of muscle alone.

Can one do massage on one's self? Not very well; for action and reaction being equal and opposite, it is somewhat like trying to pull one's self over a fence by the straps of one's boots. Moreover, if not accustomed to doing massage, one can easily ascertain the difficulty of doing this effectually and gracefully by comparing the effort with his first attempts at writing or brushing the teeth with his left hand. A young surgeon of my acquaintance undertook to *masser* his father, a surgeon of eminence. It did not take him long to rub the skin off his father's neck. In quality and endurance the strength necessary for doing massage is different from that required to scrub a floor, swing a sledge-hammer or win a boat-race. There is as much difference in massage and manipulators as there is in music and musicians, or in people in other occupations. It may be said that *savoir faire*, *gumption*, and rule-of-thumb, all go towards making a manipulator skillful. But he will doubtless at times attain this best by forgetting art and aiming beyond it in his sincere desire to do good. And this will be an instance that often "it is the heart and not the brain that to the highest doth attain," but not the one without the other. "All great art is the expression of man's delight in the work of God," but with this it should be remembered that those who are best able to resolve will be best able to combine.

Unless the patient come to the office of the *masseur* it is better that the latter should ride to the patient, in order that his pulse and respiration be not quickened. For even an acceleration of these, which is very agreeable in the open air, will make it very uncomfortable for the *masseur* to work in a closed room; and it is thus oftentimes has arisen the fallacy that he has lost vitality by imparting "magnetism" to his patient. After the *séance*, being slightly heated, he will find it easy, agreeable and salutary to walk, especially before meal time, when the products of digestion are being poured into the circulation in greatest abundance.

CHAPTER IV.

PHYSIOLOGICAL EFFECTS OF MASSAGE.

"This is an art which does mend nature,
But the art itself is nature."

THAT massage has been steadily gaining in favor with the medical profession and the public for the past ten years can hardly be denied, even though its performance has usually been left in the hands of the most common and uncommon people, to whom it is intrusted as a matter of favor, friendship, or charity, without regard to their qualities or qualifications. It is a mystery yet to be explained why patients who are proof against the most time-honored remedies and defy the most painstaking skill should be consigned to such hands, as if they were endowed with supernatural virtues. Contrary to the advice of their physicians, however, patients not infrequently prefer to fall into such hands, and, by so doing, put a premium on ignorance. Benefit or harm may follow from the roughest kind of scraping and pounding, and in a matter of such great importance as recovery from chronic and often hitherto regarded as hopeless invalidism, the means employed cannot be too carefully selected, especially when it is a question of such potent means as massage which affects, either directly or indirectly, every function of the human body.

A study of the effects of massage is, therefore, commensurate with that of physiology itself, and only a general outline of them can here be attempted. The pressure of deep massage exerts a simultaneous influence upon all the tissues within its reach, upon the skin, fasciæ, muscles, vessels, nerves, etc. The skin, by reason of its highly organized structure, is remarkably well

63

adapted for receiving and transmitting the influence of massage. Beginning at its exterior, we find that the epidermis not
only limits watery evaporation, prevents the absorption of noxious
substances, and diminishes the evolution of heat, but it also serves
as a protection to the papillary layer against the encroachment of
too vigorous friction or other external violence. The highly sensitive
and vascular papillæ on which the deeper layer of the cuticle fits
so accurately gratefully respond in agreeable sensation to judicious
friction and manipulation, or unhesitatingly complain when the
skin is pinched too strongly, or when the cuticle has suffered
abrasion. With a deeper and a more superficial plexus of nerves
whose terminal filaments register so well any morbid action at
their origin in the central nervous system, it has been reasonably
inferred that appropriate stimuli applied to them, such as massage
and electricity, often exert a favorable and curative influence upon
disturbances at their ends in the brain and spinal cord. The soothing effect of gentle stroking transmitted to the sensorium is well
known; but, somewhat to my surprise, it is not so generally
recognized that pinching the skin produces an anæsthetic effect
upon it, even to the extent of inserting a hypodermic needle without the prick being felt. The principal seat of the sense of touch,
there is perhaps no sensation that can be felt by the skin so
delightful as that arising from the contact of the hands in properly
done massage, or none so disagreeable as that from improperly
applied massage. Tough, flexible, and elastic as the skin naturally
is, owing to the white fibrous and yellow elastic tissue in its com-
position, it is rendered none the less so by a prolonged course of
massage. On the contrary, while it becomes softer, suppler, and
finer under manipulation, it at the same time becomes more tough,
flexible, and elastic, so that whereas at first it could scarcely
be gently pinched and grasped without hurting, later on, the
patient will often delight in almost being lifted up by the skin,
like one of the agile domestic animals. With the capillary blood-
vessels nearer the surface than the lymphatics, the effect of massage would naturally be greater on the former than on the latter,
and thus would be verified the remark of old Celsus, "that the
food penetrates to the skin, which has been relaxed by a kind of

digestion or removal of its tissue." Insensible perspiration, when deficient, is increased, and the sebaceous excretion is facilitated, as is best shown by the moisture and gloss of the hair after massage of the head. Weynrich has shown that by the mechanical action of friction the excretion of water through the skin can be increased sixty per cent. or more. For rousing the action of languid skin alone, friction, pinching, and percussion would commend themselves. The frequent effect of massage upon the skin and expression of countenance remote from its seat of application was well shown in a gentleman fifty years of age, of swarthy complexion, upon whom the marks of time were apparent in moderate wrinkles on his countenance, who, after a four months' course of massage for a peculiar affection of the muscles of his back, looked very much younger; his skin was clearer and smoother. One morning I said to him : "You look ten years younger, sir." He stared at me in astonishment and replied : "You are the third person who has told me that within twenty-four hours." This is one of the incidental effects of massage which is simply an expression of renewed vigor and improved general health, though in this case the general health was not previously affected, the trouble being local and not constant.

In malnutrition from digestive, respiratory and other disturbances, inunction has often proved an efficient means of furnishing nourishment to the system where other means have failed. The skin is in the best condition for absorbing oils towards the end of a *séance* of massage when its circulation has been thoroughly aroused, and for this purpose a preliminary warm bath is of great aid.

The normal function of the superficial fascia in facilitating the movement of the skin over the subjacent structures is favorably influenced by massage, especially when there exists a tough, *matted, hide-bound* condition ; and this, if looked for, will be found as often in the human race as in the equine, indicating that there is neither swiftness of motion nor clearness and vigor of thought. Its diagnosis and removal are accomplished by the same means, and the superficial vessels and nerves that pass through this fascia, besides being acted upon directly, are at the same time freed from the hindrance of pressure. Grasping a convenient portion of skin, and slowly moving and stretching it, effects this object most easily.

To aid and imitate the natural functions of the human body is often the chief aim of the physician, and, in doing this, he may constitute himself an artist of the highest order, and few, if any, remedial agents can he call to his assistance that will so exactly reproduce the mechanical forces that carry on nutrition as massage. In attempting a description of the effects of massage upon the muscles and deep fascia, and indeed upon all the other tissues, it would be desirable that the mirror should be held up to nature as perfectly as it ever has been in any work of art, and doubtless the future developments of physiology will add much to the lights and shadows of this picture and remove from it a great deal of the mysterious and unhallowed mist that has so long obscured it; for I am sure that no more fertile field awaits the investigations of physiologists than that of ascertaining the similarities and differences existing between exercise and massage. No better meed of praise could be bestowed upon any therapeutical agent whatsoever than the old-fashioned, haughty, supercilious way of dismissing the subject of massage as unworthy of notice, by saying that it was merely a substitute for exercise, and that it acted upon the mind of the patient. According to this way of reasoning, if one were deprived of air, a substitute for it, if it could be obtained, would be of no account. Appropriate exercise acts and reacts favorably upon mind and body, upon nerves and muscles, and people who can exercise freely without fatigue, and who can eat and sleep well, seldom need massage. The writer is not unmindful that this statement includes many neurasthenic individuals, especially those who suffer from want of occupation.

A study of the natural functions of the human body, alone, might teach us to use massage when they are in a state of suspension, abeyance, or morbid action. By their intermittent compression and relaxation, muscles in action exert a sort of massage upon each other. The ascent and descent of the diaphragm in respiration make continual massage and passive motion upon

the organs above and below it, more especially of the abdominal
and pelvic organs; and when its movements are limited from
want of exercise, or restrained by tight lacing, it is only too
familiar how feeble become appetite and digestion, and how
constipated the bowels become. The voluntary muscles should
receive about one-fourth of the total amount of blood in the
body, and few organs are as plentifully supplied; and their vessels may with propriety be considered as the derivative channels,
for the relief of hyperæmic conditions of internal organs. Their
action presents a great similarity to that of a beating heart, for at
every contraction of a muscle the blood is driven out of it, and by
this it at the same time receives an additional impulse in its return to the heart, while at every relaxation the vessels are again
allowed to fill. The parallel may be carried still further in order
to point out a practical lesson; for the heart, which is abundantly
supplied with blood for its own nourishment, lasts usually a lifetime without fatigue, though in constant activity, while voluntary
muscles, if allowed to remain inactive, soon suffer in size and
strength, for their "circulation goes around rather than through
them," as so aptly expressed by Dr. Weir Mitchell. Hence the
importance of some measure that will overcome the evils of inactivity, that will at once attract the circulation to the muscles and
at the same time aid in its return. This indication is, perhaps,
better fulfilled by the intermittent pressure of massage than by
any other known remedy; for it makes more blood go through
the skin and muscles, and consequently less to the brain, spinal
cord, and internal organs generally. Not that the effects of
massage and exercise are alike in all respects, and that massage is
only a substitute for exercise, as some would have it, for voluntary
exercise means exercise of the nervous system quite as much as of
the muscular, and sometimes more; besides, the cases that are
often much benefited by massage are those of overtaxed brain and
used-up nervous energy, to whom exercise, in the ordinary sense,
would only increase their exhaustion and which yet require a mechanical stimulus of their nutritive functions. True, a certain
· store of latent energy is necessary in order to undergo massage,
but this is much less than would be required for voluntary exer-

cise, were this possible. Fatigue is an indication that waste is greater than repair. Muscular fatigue from over-exertion, or want of exercise, is relieved by massage, which promotes a more rapid absorption of waste products and stimulates the tardy peripheral circulation upon which weariness to a large extent depends, thus showing a marked difference between the effects of exercise and those of massage. Fatigue from mental straining is relieved by the same means which increases the area and quantity of the circulation in the external tissues of the body, and thus depletes the over-filled cerebral vessels.

In this connection, the observations of Zabludowski on the effects of massage upon healthy people are of great importance and significance.[1] They were made in November, 1881, upon himself, then thirty years of age ; upon his servant, twenty years of age ; and upon his housekeeper, forty-seven years of age. All three were living under the same conditions with regard to food, activity, and dwelling, and, for eight days prior to the massage, examinations were carefully made of their weights, muscular strength, temperature, pulse, respiration, and urine. Observations were made during the ten days in which they had general massage, and also for eight days afterwards. The muscular strength of all three increased during massage. The weight of the one who was tolerably corpulent decreased, as also did that of the slender housekeeper, and, corresponding to this, there was an increased excretion of urates and phosphates. The weight of the one who was but moderately nourished, increased, and with this there was found a diminution of urates and an increase of sulphates in the urine. The massage of the abdomen excited the large intestine to powerful peristaltic action and caused regular evacuations. Oft-repeated observations showed that there was an elevation of the functions of life in general, and, with the improved frame of mind, there was also easier movements of the body. Appetite increased and sleep was soft, gentle, and steady. The effects of the massage disappeared soonest from the moderately nourished person, the servant ; and they lasted during the whole time of observation, for eight days after the massage, upon

[1] Zabludowski über die physiologische Bedeutung der Massage. Centralblatt für die Med. Wissenschaften, April 7th, 1883.

the housekeeper, who, though thin and slender, had lost weight, while upon Z. himself, the moderately corpulent person, the after-effects varied at different times.

Interesting observations have been made by Dr. Hopadzë showing the influence of massage on the metamorphosis and assimilation of nitrogenous food substances. He daily estimated the nitrogen of the food, fæces and urine for one week before, during one week of massage, and for a week after, and found that the nitrogenous metamorphosis in all, from persons to whom he gave daily massage of twenty-five minutes, invariably increased and lasted for seven days after the week of massage. The assimilation of the nitrogenous substances of the food increased in all the cases and lasted during the week after massage. All the four patients increased in weight during the week following the massage; but during the week of massage one gained in weight, two lost and one was unchanged. It is possible that these changes lasted longer than seven days after the massage, but the observations were not continued beyond this time. Another series of observations by Hopadzë showed that massage of the abdomen for ten minutes lessened the sojourn of the food in the stomach from fifteen to seventy-five minutes.

Salol serves a useful purpose in showing to us when the contents of the stomach pass into the small intestines. Insoluble in the gastric juice, it is decomposed into its two constituents in the small intestine, where it is subjected to the action of the contents of the duodenum, which render it soluble and easy of absorption. Soon after absorption into the circulation it is found in the urine as salicyluric acid and sulphocarbolic acid, its presence being shown by the production of a red-violet precipitate when the urine, after acidulation with hydrochloric acid and shaking with ether, is tested with a solution of perchloride of iron. The experiments of Prof. Ewald and Dr. Eccles agree in that they found in most cases under natural conditions without massage that salol could be detected in the urine in forty-five minutes after its administration, but after massage to the abdomen for fifteen minutes the reaction of salol was obtained in the urine in thirty minutes. In two obstinate cases the addition of general massage had a more favorable effect in hastening the absorption of salol than did massage of the abdomen alone. Acute catarrhal conditions and chronic dilatation delays the decomposition and absorption of salol.

After the administration of one gramme of salol to people suffering from chronic dyspepsia, Hirschburg, Brunner and Huber found that it required from two hours to two and a half hours before the reaction of salol could be detected in the urine. After a walk of fifteen minutes or gymnastic exercises for ten minutes the reaction was obtained in one hour and five minutes. Similar results were obtained from faradization, but massage proved more efficacious than any other means in hastening the passage of salol from the stomach to the duodenum.

Zabludowski has also made some interesting experiments to learn how fatigued muscles are influenced by massage. Muscles of uninjured frogs were exhausted by a series of rhythmic contractions caused by an induction current. Under massage they soon regained their lost vigor, so that the contractions were almost equal to the first, whilst a rest for the same period, without massage, had no effect. These experiments, showing the restorative effects of massage upon wearied muscles, were more than confirmed in man by the same investigator. He found that after severe exercise a rest of fifteen minutes brought about no essential recovery, whilst after massage for the same period, the exercise was more then doubled. One person experimented upon lifted a weight of 1 kilo (2.2 pounds) 840 times, at intervals of one second, by extreme flexion of the elbow-joint, from a table upon which the forearm rested horizontally, and after this he could do no more. When the arm had been *masséed* for five minutes, he lifted the weight more than 1100 times in the same manner as before without fatigue. The difference in muscular sensation was very striking after rest alone from work, in comparison with that after massage. In this case, the person experimented upon was an expert subject for experiment, and after he had made 600 lifts of 2 kilos (probably in the same manner as that just referred to), there was unvarying stiffness during a pause of five minutes for rest; on the contrary, after five minutes' massage the muscles felt supple and pliant.

Of the influence of massage on reflex irritability Zabludowski found by a series of observations on rabbits, that sensibility was lessened whilst the reflex action of the spinal chord remained without change.

Later observations also showed him that when rabbit-muscle had been tetanized by means of an induction current, the motility of the muscle was only imperfectly and transitorily restored after ten minutes of rest, so that they could be very easily again thrown into a tetanic condition. But when, in place of rest, ten minutes of massage was used, the power of motion of the muscles became four or five times longer than before.

Kroneker and Stirling have shown that muscles, when fatigued, can be tetanized by much less frequent irritation than when fresh and rested. A fresh muscle that receives six irritations per second passes gradually from its intermittent contraction into that of tetanic contraction. Later, upon fatigue, this will be less. If the muscles are allowed to recover by rest alone for a short time, upon renewed irritation they very soon pass into a tetanic condition. If, however, during the same pause for rest the muscles have been *masséed*, then their motility returns, so that they have the power of contracting a great many times, often 100. According to this, massage is considered by these observers to act as a perfect *perfusion*, bringing nourishment to the muscles and thoroughly removing asphyxiated juices from them. It was found that the sensitive nerves of the skin lost considerable of their irritability during massage; but over-irritation, in consequence of strong rubbing, might sometimes be made available.

Upon testing the influence of massage over the irritability of muscles, — their power of receiving an impression in contradistinction to their capacity for action, — the unexpected result was obtained that irritability was diminished by massage. In one experiment, after a long series of contractions caused by the secondary current, when this stimulus had become ineffectual in consequence of fatigue, the muscles recovered tolerably well after rest alone for twenty minutes, so that the same intensity of current was again effectual. When after repeated fatigue from the same cause the muscles were *masséed* during rest, the current had to be increased before the muscles could be made to contract.

My own observations, repeated almost daily, teach me that
muscles give a much more ready, vigorous, and agreeable response
to the will and to the faradic current after massage than they do
before, especially if they are somewhat deficient in contractility.
But in my cases the trouble is not taken to fatigue the muscles by
electrical stimuli before massage, but only to test them briefly.
We may therefore conclude that massage lessens irritability, but
increases power of action.

If we first call to mind the great extent of ramifying tubes and
cavities formed by the deep fascia and investing membranes, we
will more fully realize the importance and significance of the function of the lymph spaces in fasciæ presently to be spoken of.
Besides enveloping the four hundred muscles of our bodies collectively, the fascia surrounds each by a separate sheath, and aids
them in their action by its tension and pressure upon their surface. In the limbs it gives off septa which separate the various
muscles, and are attached to the periosteum beneath. It also
forms sheaths for the innumerable vessels and nerves ; and if all
the organs of the body could be removed without injury to their
investing membranes, there would still remain an exact outline
of their form and position, and the whole body (or rather what
would be left of it) would present a skeleton of cavities and
tubes beautifully adapted for the support and protection of the
various organs. From a paper by Professor H. P. Bowditch,
published in the proceedings of the American Academy of Arts
and Sciences, on the "Lymph Spaces in Fasciæ, with a New
Method of Injection," we learn the following valuable facts:
"The lymph spaces existing between the tendinous fibres of
fasciæ and the connection of these spaces with lymphatic vessels,
have been well described by Ludwig and others. The researches
of Genersich have shown that the fasciæ, by virtue of this structure, play a very important part in keeping up the flow of lymph
through the lymphatic vessels. A piece of fascia was removed
from the leg of a dog, and tied over the mouth of a glass funnel
with the side next the muscles uppermost. A few drops of a
colored turpentine solution were then placed upon this surface,
and the fascia alternately stretched and relaxed by partially ex-

hausting the air from the funnel and allowing it to return again. In this way the colored matter was made to penetrate into the spaces between the fibres of the fascia and to enter the lymph spaces upon the opposite side. The same result was obtained when the coloring matter was injected between the muscles and the fascia, and the latter stretched and relaxed by passive movements of the limb. Experiments on animals, where the flow of lymph through the thoracic duct was measured, showed that passive movements increased this flow in a striking manner. Galvanization of the muscles had a similar but less striking effect. The alternate widening and narrowing of the lymph-spaces between the tendinous fibres seemed, therefore, to cause absorption of the lymph from the neighboring parts as well as its onward flow into the lymphatic vessels, the valves in these latter preventing a flow in the opposite direction." This generally overlooked function of the fascia certainly affords a partial, but important, and, so far as it goes, very satisfactory explanation of the success of methods of treatment involving passive movements; for the removal of worn-out matters from the tissues is undoubtedly favored by an increased flow of lymph.

Here we have one of the secrets of nature revealed to us, and one which she is continually performing as one of her regular functions, by means of voluntary and involuntary muscular action. But in admiration of this comparatively recent discovery of the function of lymph-spaces in fasciæ, we must not forget that the lymphatics are still more abundant in organs that are well supplied with blood-vessels, such as the muscles. The meshes of the lymphatic plexus being interposed between those of the capillary blood-vessels, in the transudation of the nutritive fluids from the latter to the former the intervening tissue is completely traversed before passing through the point of junction of two or more lymphatics in the middle of the space surrounded by the adjacent blood-capillaries. But the removal of effete matters from the tissues is not the only function of the lymphatics, for while the blood is being returned to the heart and lungs by the veins, the lymphatics take up more slowly the fluids which have served for nutrition and growth, and also the superabundance of nutritive fluid not immediately required for the nourishment of the tissues. The lymphatics in accomplishing this, besides their primary and peculiar function of endosmosis, are materially aided in their absorptive power and centripetal movement of lymph by the pressure of the blood, by the natural elasticity of the tissues, and by the contraction and relaxation of the muscles. Now all these forces can be increased to a much higher degree by the externally applied pressure of massage which, being intermittent, does not hinder the circulation. The pressure of a fluid from endosmotic action alone can support a column of mercury at the height of six hundred millimeters (twenty-four inches), and soluble substances that will not ordinarily transude may be made to do so by increasing their pressure or rapidity of movement. Repeated measurements have shown me that eight or ten minutes of deep massage on a leg of ordinary size caused a temporary decrease in the circumference of the calf of one-fourth of an inch. If massage be continued until a markedly hyperæmic condition results, it is doubtful if the limb more than resumes its former size at the same sitting, even when in the intervals between the applications increased growth is going on. Lassar found that massage of the lymphatic glands, whether healthy or inflamed, caused large quantities of lymph to escape from them, but electrical irritation had no such effect.

The peritoneal and pleural cavities, and probably all the other serous sacs, are in communication with the lymphatic vessels. The cavities of the pleura and peritoneum are now regarded as extensive lacunæ in the course of the lymphatic vessels; lymph-spaces and lymphatic vessels communicating with each other by means of small openings, or stomata, have been demonstrated in these membranes, and also the communication of the lymph spaces with the pleural and peritoneal cavities by means of intercellular openings. This has been shown by injecting these cavities with colored fluid, and after killing the animal, examining the course of absorption of the fluid under the microscope. A drop of milk placed on the peritoneal surface of the central tendon of the diaphragm which had just been removed from a recently-killed animal, was seen under the microscope running in convergent currents to certain points on the surface of the tendon, and thence the milk-globules were observed penetrating into the lymphatic vessels. In the respiratory movements of alternate expansion and contraction of the chest-walls, with descent and ascent of the diaphragm, we have a continual pump-like action of absorption and onward expulsion in the lymph-spaces and lymphatic vessels of the pleura and peritoneum as well as in those of the muscles and fasciæ of the chest and abdomen. It will now be evident why the kings of the Sandwich Islands have had themselves *lomi-lomied* after every meal as a means of aiding their digestion, for the externally applied intermittent pressure over the abdomen would force the contents of the lacteals, or lymphatics of the small intestine, onward, at the same time aiding them in their absorption of digestive products. A similar effect would be exerted upon the blood-vessels of the intestinal villi.

Reibmayr inserted a small glass tube into the lymphatic vessel which accompanies the saphenous vein of a dog, and found that no flow of lymph took place through this so long as the leg was quiet. But as soon as the paw was moved, or muscular contractions were excited, lymph flowed freely from the tube. Centripetal stroking or kneading of the paw, although this was at rest, had the same effect. At first the flow was abundant, and then gradually diminished, and after a short interval increased again. Lassar had similar experience with the paws of dogs in which inflammation had been artificially produced. When the inflamed leg was manipulated, or passively moved, lymph flowed abundantly from the divided absorbents, and this was much greater in quantity than that obtained from a sound leg of the same animal by like procedures. The flow was seven or eight times more plentiful than that from the sound leg, and in the latter it was only obtained with much greater efforts of kneading and passive motion The inflamed and swollen extremity diminished in circumference. and finally the flow ceased altogether. Considerable time elapsed, before lymph could be again obtained in this manner. From these experiments may be drawn the inference that there is a limit which must not be overreached in practice, the sensations of the patient and the state of the affected parts often indicating that the limit of temporary benefit has been obtained.

Not only is the effect of massage upon the rootlets of the

lymphatics in the fasciæ and muscles of the greatest interest, but of quite as much importance is its influence also on the large lymph cavities accessible to its intermittent compression, such as the peritoneal cavity, the synovial cavities, and the cavities of the sheaths of the tendons.

The experiments of Reibmayr and Hoffinger, showing that the absorbtive power of the peritoneum is increased by massage, support clinical observations, and are regarded as conclusive. Measured quantities of water were injected into the peritoneal cavities of rabbits. The animals were killed at the end of one and two hours respectively, and the quantity of fluid remaining was ascertained in each case without any massage being used. The same quantities of water were injected into the peritoneal cavities of other rabbits, and their abdomens were kneaded for a short time every fifteen minutes. Some of the animals were killed at the end of an hour, others at the end of two hours, and the fluid remaining in the peritoneal cavity of each was accurately measured. Though the natural absorbent power of the peritoneum is very great, yet under the influence of massage twice as much fluid was absorbed during the first hour as there had been without massage. During the second hour, on the contrary, only half as much fluid was absorbed under massage as without; but notwithstanding this, there was so much more absorbed altogether during the two hours in which massage was used, that the total amounted to 39 per cent. more than the total without massage. The proportion absorbed without massage to that with massage was as 7.40 to 10.29, the difference, 2.89, being a fraction over 39 per cent. more than without massage. (Reibmayr has made an error in his own figures in stating this difference, 2.89, as per cent.[1] In calculating the proportion of fluid absorbed to that of the weight of the animal, Reibmayr has made another error in not distinguishing any difference between grammes and centigrammes. Thus in experiment No. iv. he states that 105 centigrammes were absorbed, and that this is 10.29 per cent. of the weight of the animal, 1,023 grammes; just 100 times too much. He has applied the same method of calculation to all his other experiments.)

_____

[1] See page 33 of Reibmayr on Massage.

These experiments would again point out the lesson that there is a limit to the benefit to be derived from a single *séance* of massage. beyond which it is useless to prolong it. But these observers, to whom we are already so much indebted, should have made a third series of experiments showing the quantity of fluid that might have been absorbed during an hour of repose, following an hour of massage, to compare with the results of two hours with and without massage.

Evidence in favor of free circulation would be almost as superfluous as was Sancho Panza's apostrophe to sleep. Running water frees itself from impurities to a great extent, and unimpeded circulation may well be likened to a running stream, doing useful work and keeping the machinery of the various districts of the body in motion and in health. A rapid flow, whether in a river, blood-vessel or lymphatic, prevents the deposit of particles held in suspension. In an interesting article on the " Influence of Rest and Motion as the Phenomena of Life " (*Archiv für die Gesammte Physiologie*, Vol. 17, page 125), by Dr. Horvath, of Kieff, we are told that it was with the greatest astonishment that the observer saw that the direct introduction of bacteria into the arteries, instead of increasing or calling forth disease, on the contrary produced no observable indisposition in the animals so treated. Indeed, the bacteria disappeared to such an extent that they could not be found in the blood of the vessels where they had been injected in the greatest abundance. Animals that die from bacteria poisoning have the bacteria for the most part in the lymphatics, never in the large arteries, where we might expect to find them, for it is here they get the oxygen requisite for their development, present in greater abundance than in the veins and lymphatics.

For further evidence of free circulation overcoming the influence of noxious substances, we have only to recall the fact that laborers will work in an open sewer with impunity, while business or professional in their offices near by would be made sick if they should leave their windows open. Let the circumstances be reversed, and those in active exercise would be less likely to suffer than the sedentary laborers. It has been estimated that a person in exercise consumes four or five times as

much oxygen as he does when at rest, but in the case of one
working in a sewer, the air is so vitiated that we must conclude
that it is not the oxygen, but the exercise and the active circulation that keep off the bad effects of the effluvia. Even local
stagnation of the blood from injury or other cause may lead to
pathological changes resulting in death, or requiring the ampu-
tation of a limb.

It is, indeed, a wonder how the lymphatic and venous currents ever do get back to the heart, so far removed are they from
its propelling influence and with so little else to aid their return.
Nature, forseeing the disadvantages under which the returning
circulation labors, has wisely made the capacity of the veins
double that of the arteries, and strengthened their coats by a
greater abundance of condensed connective tissue in order that
they may be prepared to stand a greater strain from pressure than
the arteries. Still doubting her work, she has supplemented the
veins with the lymphatics, but for all that, when the contraction
and relaxation of the voluntary muscles no longer take place, the
returning circulation languishes, and consequently the outgoing
as well. Now it is that the assistance of massage becomes in-
valuable, for by upward and oval friction with deep manipulation, the veins and lymphatics are mechanically emptied, the
blood and lymph are pushed along more quickly by the additional *vis a tergo* of the massage, and these fluids cannot return
by reason of the valvular folds on the internal coats of their vessels. More space is thus created for the returning currents
coming from beyond the region *masséed*, and the suction power
induced at the same time adds another accelerating force to the
more distal circulation. In brief the effect may well be likened
to the combined influence of a suction and force-pump, and in
people that are not too fat the superficial veins can be seen collapsing and filling up again as their contents are pushed along
by the hand of the *masseur*. In this way the collateral circulation in the deeper vessels is aided and relieved, as well as the
more distal stream in the capillaries and arterioles. One would
naturally suppose that the circulation in the larger arteries
would, in this manner, be interrupted, and such is the case.

But herein comes an additional advantage to aid the circulation, for the temporary and momentary intermittent compression causes a dilation of the arteries from an increased volume of blood above the parts pressed upon, and this accumulation rushes onward with greater force and rapidity as soon as the pressure is removed, into the partially emptied continuation of the arteries, in consequence of the force of the heart's action and the resiliency of the arteries acting upon the accumulated volume of blood. But the same pressure, as we have seen, also acts upon the tissues external to the blood-vessels, causing a more rapid absorption of natural, and also of pathological products through the walls of the lymphatics and venous capillaries.

The apparent mystery and contradiction of many physiological experiments need seldom arise, if it only be borne in mind that irritation, when mild, produces symptoms of stimulation; when stronger or longer continued, symptoms of exhaustion. Thus gentle centripetal stroking, though soothing, is, in a physiological sense, a mild irritant of the superficial vessels, causing a narrowing of their calibre and a stronger and swifter current in them by reason of its stimulating influence on their muscular coat and vaso-motor nerves. But let centripetal stroking, or any other form of massage, be continued sufficiently long, or become stronger, and hyperœmia will result, indicating relaxation of the vascular walls due to over-excitation or exhaustion of the tone of their muscular coat and vaso-motor nerves. But retardation is obviated by the mechanical effect of the massage pushing along the returning currents, so that the ultimate effect in either case is an increased rapidity of the circulation.

It will now be evident that massage rouses dormant capillaries, increases the area and speed of the circulation, furthers absorption, and stimulates the vaso-motor nerves, all of which are aids and not hindrances to the heart's action and to nutrition in general. Seeing that more blood passes through regions *masséed* in a given time, there will be an increase in the interchange between the blood and the tissues, and thus the work done by the circulation will be greater, and the share borne by each quantity less.

Exercise accelerates the action of the heart and diminishes blood-pressure, which means an increase in the rapidity of the current, and in the quantity of the flow through the relaxed, distended, or stretched blood-vessels. Massage also diminishes blood-pressure, but without increasing the activity of the heart. On the contrary, the heart's action is generally lessened in force and frequency. And, on reflection, this is what might be expected; for natural obstacles to the circulation are gravity and the friction of the blood against the walls of the vessels, and these working backwards to the heart, have to be overcome at each systole of the left ventricle. These hindrances are by massage both directly and through the medium of the vaso-motor nerves in great part removed. The contracting hands of the manipulator are, as it were, two more propelling hearts at the peripheral ends of the circulation, co-operating with the one at the centre, and the analogy will not suffer if we bear in mind that the size of one's heart is about as large as the shut hand, and the number of intermittent squeezes of massage that act most favorably on vessels, muscles, and nerves are about seventy-two per minute, which is about the ordinary pulse rate. If this is not an art that does mend nature, what is?

But the walls of the blood-vessels possess an intrinsic tone of their own, whether dependent on some local nervous mechanism or not. The muscular walls of the vessels, like those of the intestines, are composed, as we know, of involuntary muscular fibres which respond by contracting very slowly on the application of stimuli, mechanical or other. Voluntary muscles, for a short time after death or removal from the body, will contract readily on being pinched, percussed, or galvanized, and when they cease to respond they can be temporarily restored by injecting fresh arterial blood through them. Involuntary muscles can also be made to contract in their peculiar, slow, vermicular manner on the application of stimuli after their removal from the body, as in a piece of intestine. During life, of course, they respond much better to the same stimuli, mechanical, chemical, electrical, or thermal. But muscles cannot contract well unless they can also relax well. Tension or extension of a muscle within natural limits increases its power of contraction. Life is made up of a series of activities and passivities alternating with each other. A heart well filled, and thus relaxed, beats more strongly than one but partially filled, and so distention of the intestines increases peristaltic action. By the stretching and pressure of massage we obtain and increase extension or distention and imitate and stimulate the alternating contraction of voluntary and involuntary muscles that are accessible, and also of those that are inaccessible, by sympathy and by reflex action. Proof of this is seen in flabby and relaxed abdominal muscles gaining in tone and firmness even while being *masséed*, and in the regulated and more vigorous action of the intestines, especially of the large intestine. So undoubtedly the same improvement in turn takes place in the vascular system, particularly where muscular fibres are found, as in the veins, lymphatics, and smaller arteries.

Physiological experiments which involve section or destruction of tissue are usually of too pathological a character to present a trustworthy analogy between themselves and what takes place in the uninjured body under similar circumstances. But when they agree with clinical observation and common sense they may be accepted as corroborative testimony. Thus we would judge of the experiments of Golz (Virchow's Archiv, Bd. XXVIII., page 428), in which, after opening the abdomen of an animal and applying percussion to the stomach and intestines, the peritoneum at first became paler from constriction of the vessels, but on continuing the percussion this was soon replaced by dilatation. When the percussion was first applied thoroughly over the abdominal walls, and these laid open afterwards it was observed that the vessels of the abdominal cavity, especially the veins, were dilated and distended with blood. The distention was due to relaxation of the vascular walls caused by the mechanical irritation, and this might have been increased to the extent of paralyzing them. The heart's action was materially retarded owing to the reflex influence of the percussion upon the inhibitory action of the vagus, and also to the sudden withdrawal of blood into the abdominal vessels. The pulsations of the heart gradually became less and finally ceased. Respiration also became less frequent and finally ceased, and symptoms of motor paralysis were induced in like manner. In practice the lesson drawn from these experiments would be, that percussion briefly applied may be used to cause vascular contraction, longer continued to induce dilatation.

The experiments of Bernard on dogs gave results similar to those of Dr. Beaumont on the inner coat of the stomach of Alexis St. Martin. When the mucous membrane was gently stroked with a glass rod, the natural pale pink color became rosy red and secreted juice abundantly ; but when violently rubbed the color disappeared and became pale, the secretion of gastric juice stopped, the mucus increased, sickness and vomiting followed. The primary dilatation of the vessels caused by gentle stimulation seemed to be replaced by contraction under more violent irritation, the opposite of what usually occurs out of the stomach.

Upon the nervous system as a whole, massage most generally exerts a peculiarly delightful, and at the same time profoundly sedative and tonic effect. While it is being done, and often for hours afterwards, the subjects are in a blissful state of repose ; they feel as if they were enjoying a long rest, or as if they had just returned from a refreshing vacation, and quite frequently it makes optimists of them for the time being. An aptitude for rest or work usually follows, though generally those who submit to this treatment feel gloriously indifferent, and needless apprehensions are dispelled. With much less expenditure of time and money a course of massage at home serves many much better than a vacation with anxiety about business. I have never known anyone to take cold or suffer from exercise in the open air after general massage when ordinary care was observed. An able writer in the *British Journal of Mental Sciences* recommends massage " for certain melancholics with trophic and vaso-motor affections, and where dementia is threatened after an attack of excitement. Under this treatment, mental comfort and a sense of well-being take the place of apathy and lassitude." Through the medium of the central nervous system, even local massage is radiated or reflected throughout the body, thus acting at the same time as a nervous and vascular revulsive, or physiological counter-irritant, if one may be allowed this expression. One of the best examples of this, perhaps, is the relief of headache from the manipulation of the back and shoulders. It has long been well known that stroking the limbs often induces sleep. Massage of one part of the body is sometimes accompanied with a peculiar, but not disagreeable sensation of tingling or crawling in some other part. In one person this was experienced on the outside of the thigh while the corresponding side of the head was being *masséed;* and in another in one leg while the other was being acted upon. In two persons I have witnessed an approach to syncope from the first attempt at massage of the head. Such peculiarities must be very rare, and also those idiosyncrasies in which massage cannot be tolerated at all when it is apparently indicated. A more frequent concomitant of massage is that of an agreeable thrill such as we are apt to experience on the receipt of joyful news, or on learning of heroic actions. Morbid irritations are reflected in a similar manner, but with different effect. We all know the general depressing effect of local pain, and we also know that neuralgia or other trouble sometimes appears distant from the original seat of disturbance, and in no way connected therewith, except through the cerebro-spinal axis. The transmitted and reflected influences of massage must evidently be as numerous as the distributions and connections of the sensitive nerves that are accessible to its impression. Briefly it may be said to act on distant parts by sympathy, by reflex action, and by inhibition. In Dr. Weir Mitchell's cases, which gained flesh under massage, rest, and feeding, the gain was first noticed in the face. This precursor of increasing nutrition may have been owing to a remote influence of massage, for the faces were not *masséed.*[1] Doubtless due allowance was made for the fact that most people, on assuming the horizontal position, even for a brief period, look fuller in the face, owing to the effect of gravity upon the circulation.

This brings us to a consideration of the immediate effects of massage upon nerves, for on their integrity depends the perfection of their functions. The immediate effect of massage by gentle stroking in producing a soothing sensation upon the seat of ac-

[1] Brown-Séquard states that he has produced flushing of the auricle by pinching the sciatic nerve in his own person. St. John Roosa on the Ear, p. 112.

tion, and the anæsthetic effect of vigorous pinching have already been referred to. Percussion is a good means of exciting languid nerves, and it may be continued sufficiently long and vigorously as to overexcite them, thus wearing out their capability of perceiving impressions and allaying morbid irritability. But massage may be exerting a favorable influence upon nutrition in general, while the patient is totally indifferent as to the usual agreeable sensation of its application, and why ? Because undoubtedly the nerves are insufficiently nourished, and may continue so in spite of suitable food and tonics and ordinary exercise, until their languid circulation is aroused by massage. This has the same effect upon the vessels of nerves that it has upon those of muscles, and ultimately, though not so soon as in the case of muscles, the same result is obtained, improved nutrition and with this improved function. When we call to mind that the essential element of nerve-fibres, the axis cylinder, is a delicate soft, solid, albuminoid substance possessed of a certain degree of elasticity, we can readily see that the alternate contraction and relaxation of voluntary muscles upon nerves passing through, under, or between them must be of considerable importance in keeping up their normal tone. In the absence of this mechanical stimulus, a still greater of a similar kind can be made to take its place by the intermittent pressure of massage, which should be of a uniform character in its imitation of muscular action, and this will be so much greater as the increase of pressure made by massage is greater than the compression that the muscles in contracting make upon each other, plus the influence of the compression of massage upon the cutaneous, subcutaneous, and other accessible nerves not directly squeezed by muscular contraction, and all this while the nerves are at rest, which is still another advantage. If nerve force cannot give itself expression in motion, there may be, and often is, too much left that goes to the account of sensation, and in this way the balance is often upset, giving rise to sensory disturbances. The writer of this, in an article published in 1874, has said : It is not desirable to consider massage, or any other method of treatment, a panacea, but it is evident how extensive its usefulness may become when we reflect upon the number of maladies that doom patients to a long period of inertia, the result of which is often to cause a preponderance of sensory phenomena, usually styled "nervousness," for lack of some treatment like massage to preserve the natural irritability and use of the muscles, and thus allow the motor nerves to give expression to the excitations to which they are subject in common with sensitive filaments. This feeble attempt at prophecy has been rather more than fulfilled. There are those who have overtaxed their motor functions, and others still who have exhausted motor, sensory, and intellectual powers, and need absolute rest, at least for sufficient time to give their recuperative force a chance to assert itself before massage should be employed.

It is supposed, but not proven, that in life the axis-cylinder of nerve-fibres is in a fluid condition, and that the transmission of sensations and the impulse for movements are carried along by literal wave-like movements. If this were true, the immediate and remote effects of massage would be easily understood. Some people, having in mind the undulatory theory of the transmission of light and sound, speak of sensation and motion as travelling along the nerves in a similar manner, by vibrations; and the modification of these by massage and percussion they put down positively as the way in which benefit results from this treatment. Certainly it is not the only way. When an excitation takes place in a nerve, a change in its electric state is said to occur, and assuming the axis-cylinder to be made up of electric molecules, it is imagined by Du Bois Reymond, Ludwig, Virchow, and others, that every two of these molecules take up an altered position with regard to one another at the moment the stimulus is applied. But this cannot be seen, except on paper and with the mind's eye. Excitement in all nerve-fibres is capable of transmission in both directions, and when action occurs only at one end, it is because there is a terminal apparatus capable of expressing the action present at that end. Hence we have the influence of massage passing in both directions at once, whether owing to a change in assumed electrical molecules or not. But the molecular forces, cohesion, adhesion, and chemical affinity, prevail in the tissues of the body as well as out of it, and the beneficial effects of massage are, no doubt, in great part due to its influence in bringing about molecular changes in the nerves and muscles and tissues generally, by which the chemical combinations that form the bases of the body are favored, the separation and elimination of others hastened, and greater activity and better equilibrium of the vital forces promoted. Respiration is deepened and more prolonged, the stretching of the air-cells by inspiration, and the passage of carbonic acid and oxygen through them stimulates the terminal filaments of the pneumogastric nerve, this stimulus is conveyed to the medulla oblongata, the respiratory centre is excited, which in turn sends motor impulses to the phrenic, intercostal, inferior laryngeal, and other nerves. The diminution of the respiratory movements, the lessening of blood-pressure, and the increase of the normal irritability of the muscles all correspond to an increase of oxygen and a lessening of carbonic acid in the system. These also are in harmony with the diminished action of the heart obtained by massage, for the cardio-inhibitory centre is affected by sympathy with the neighboring respiratory centre in the medulla oblongata, as is also undoubtedly the vaso-motor centre.

In a patient who had been a great pedestrian, and who suffered from severe and long continued pains in the calves of his legs, Dr. W. W. Keen found an albuminuria of from three to fifteen per cent., which disappeared quickly on resting, but reappeared promptly on resuming walking. On examining the urine immediately before and after the patient had submitted to massage for forty or fifty minutes, no trace of albumen was found. This would indicate the promotion of nutrition without such changes of blood-pressure and vaso-motor tones as induced the albuminuria after voluntary exercise.

Edelfsen noticed a transient albuminuria after exertion in three healthy but anæmic men. Leube examined the urine of a number of healthy soldiers in the morning, and found it normal; but after a five hours' march in warm weather one per cent. of albumen was found in sixteen per cent. of the cases, but no casts or blood-corpuscles.

The results of Tigerstedt's experiments on the action of mild forms of extension showed that the irritability of nerves increased under moderate extension, but grow less when the extension was increased beyond a certain limit. These results were constant and satisfactory. Zederbaum has demonstrated that when sudden and heavy pressure is applied to nerves their irritability is rapidly lessened; but when the same pressure is gradually increased up to the same extent, the decrease in irritability is not so marked and occurs more slowly. It has been found by Luderitz that motor nerve-fibres are more easily paralyzed by continuous pressure than sensitive ones. The quality of percussion, whether light or strong, seems to have an effect upon nerves similar to pressure applied in like manner. Light percussion increased the irritability of nerves, while slow and strong percussion exhausted them. Quickly repeated percussion increased the contractility of muscles supplied by the nerve operated upon, but if kept up for a comparatively long time exhaustion of the nerve resulted.[1]

Upon sensitive nerves percussion produces effects similar to what it does on motor nerves: when lightly applied at first there is an increase of pain which soon diminishes, then disappears altogether and gives place to complete loss of feeling. The more sensitive the nerve the less force and time are required to bring about these changes. Now this is quite analogous to the effects of percussion on the vascular system, which at first causes contraction and later dilatation, and if longer continued symptoms of paralysis of the vascular walls. This is all in harmony with my statement that irritation when mild produces symptoms of stimulation, when longer continued or increased in severity, symptoms of exhaustion.

I have found that the points that give the best contraction to percussion are also the same points that give the best contraction to the faradic current, and it is often surprising how much better contraction may be obtained from percussion than from a faradic current.

It is a wonder that greater thermal changes are not produced by massage; for arrested force, as friction, percussion, and compression, develop heat in the body as well as outside of it. Cool-

---

[1] Dr. Geo. N. Jacoby in Journal of Nervous and Mental Diseases, April, 1885.

ing by radiation and insensible perspiration lessen these so that they do not correspond to the usual comfortable warmth and glow that patients experience from massage. Elevation of temperature is often probably greater with the manipulator than it is with the patient, owing to an unnecessary expenditure of force.[1] The effects of vigorous exercise on temperature have, to my knowledge, never been thoroughly studied any more than the action of alcohol on the circulation.

With those in whom the temperature is normal a change is not likely to occur from massage, but if the temperature be a degree or so below normal, a rise to normal or very near will usually follow. In nervous or hysterical women with high or low temperatures, Dr. Weir Mitchell often noticed at first a slight fall in the thermometer, then a fairly constant rise, and as health improved, less marked changes. He found the most noticeable rise in those who had some organic disease and a natural liability to great changes of temperature. Zabludowski found that the perception of temperature was lessened by massage, though at first it was slightly increased. My own experience has shown me that patients are much more indifferent to extremes of heat and cold after massage than they were before.

By a series of careful and elaborate observations on healthy people Dr. Symons Eccles found that under the influence of muscle-kneading the axillary and surface temperature invariably rose while the rectal temperature fell, — the axillary rose on an average 1.4°, the rectal fell .8°. The opposite effect was produced by kneading the abdomen. When this was done immediately after the general muscle-kneading the axillary temperature fell. 2° and the rectal gained 2.2° under thirty minutes of abdominal massage, which made a difference of .6° less for the axilla and an increase of 1.4° for the rectum from what it was before the general massage.

General muscle-kneading caused a variation in the pulse rate

---

[1] When there is a marked elevation of temperature of 6° to 10° F. from massage, as in the case of children suffering from essential paralysis, Dr. Mitchell attributes part of this rise to the contact of the warm hand, and this certainly is very much warmer while doing massage.

from a diminution in most cases to an increase of twenty beats a minute in a few. But in these cases, whether the pulse was accelerated or retarded, there was always an increase of blood-pressure, even when the skin was red and warm and the vessels dilated, due, it seemed to be, to increase in the force of the action of the heart. Stroking or effleurage alone increased the pulse rate.

Massage may be exerting a beneficial or harmful influence when neither pulse nor respiration gives any indication of it, but there are other means of judging.

Under the direction of Prof. Wm. Winternitz of Vienna, Dr. Otto Pospischil has made some calorimetrical studies, and amongst these some showing that mechanical irritation, as by friction and rubbing the skin, increases the heat-loss about ninety-five per cent. Hence the value of friction of the skin in fever with excessive retention of heat. In other words, we suppose this means that cooling by radiation is favored. And thus would be verified the saying of Celsus, that "a patient is in a bad state when the exterior of the body is cold, the interior hot with thirst; but, indeed, also the only safeguard lies in rubbing, and if it shall have called forth the heat into the skin it may make room for some medicinal treatment."

Dr. P. O. Glovetzky has made a long series of experiments on man and dogs to elucidate the influence of abdominal massage on the circulation and respiration. He found that the upper half of the body increased in weight after the séance as shown by Mosso's balance, and plethysmographic measurements proved that during the sitting both the upper and lower extremities increased in volume, but returned to their normal size after massage. The upper extremities more often showed a consecutive rise in their bulk while the lower were found not infrequently to have de-creased in volume. The blood tension and intra-cranial pressure invariably rose during the massage, and the elevation lasted for a certain period afterwards. The pulse at first became quick and smaller, but towards the end of a séance it was full and slow. Respiration became more energetic, and in artificial asphyxia the cardiac action was benefited.[1]

---

[1] Vratch No. III., 1889.

The effects of massage in changing the response of muscles to electricity has already been mentioned. But further observations in this and other directions are necessary in order to show in what ways the natural or inherent electricity of the tissues is altered by means of massage, whether currents are changed in strength or direction so as to be beneficial or not, or whether the alteration of the natural currents may only serve as a favorable or unfavorable indication. It can be no longer doubted that changes of some kind do occur, for decomposition of natural electricity takes place whenever bodies are subjected to disturbances of any kind, be it friction, percussion, heat or chemical action. The electrical properties of tissues are regarded as in direct proportion to the activity with which changes of matter proceed in them, and as massage produces rapid nutritive changes, it would be expected to proportionately alter the electrical properties of tissues. Even dissimilar stretching of the skin has been shown to give rise to electromotive action. The palm of the hand has been found to be negative to the back of the hand, the whole hand negative to the elbow, to the chest and usually to the foot; and currents are found to pass from a longitudinal surface to a transverse section of muscle, nerve or brain; and during the contraction of a muscle or the activity of a nerve, the natural current diminishes. But while there is no lack of evidence to show that currents are constantly passing in and between the tissues of the same individual, there is as yet no proof that electricity can be transmitted from one person to another. Those who talk so much about imparting their own electricity or magnetism are usually too ignorant to comprehend the one experiment that comes nearest affording them grounds for their assertion. Thus, Prof. Rosenthal has shown that the power of the will can generate an electric current and set the magnetic needle in motion, simply by contracting the muscles of one arm while the other is at rest. The current then ascends the contracting arm and goes to the passive one, and this may be reversed if the muscles of the passive arm be contracted while the other rests.

The manner of doing massage which I have described, one

hand contracting as the other relaxes, would, as shown by this experiment, give the patient the benefit of the doubt as to whether a to-and-fro current could, in this way, be made to traverse the tissues manipulated, and thus combine a utilization of force with ease and efficacy of movement. Certainly, the way in which the so-called *magnetic doctors* work is not in harmony with this experiment, or the way of doing massage here described. But still the benefit, if any, that might be derived from this source, would be so little in comparison to the other effects of massage, that it would be much better to take at once a definite quantity and intensity of electricity directly from a battery.

Hysterical women may, but are not likely to, be put into an hypnotic state by means of massage, just as they might in other ways, by gentle prolonged stimulation of the sensory nerves of the face, or of the optic or auditory nerve. Heidenhain considers that in this condition there is inhibition of the activity of the ganglion-cells of the cerebral cortex, and that this is not due to anœmia reflexly produced by contraction of the cerebral vessels, for he induced hypnotism in one person during the action of nitrite of amyl, which dilates the blood-vessels.

Further discussion of the physiological aspects of massage might be indulged in without exhausting them. It will not be forgotten that "the limits of science are like the horizon, the more we approach them, the more they seem to recede from us," and that a lifetime is considered too brief to investigate the action of a single drug. The same remark may be applied to massage which that able therapeutist, Prof. H. C. Wood, has made about medicines : " In what way medicines produce changes in the life-actions of various parts is, and probably must ever remain, unknown, precisely as it is beyond the limit of human intellect to know why the nerve-cell or the spermatozoon performs the prodigies of which it is capable." But discuss any therapeutic agent as we may, there is something still peculiar to each that evades expression by tongue or pen. Of what use is it to describe odors, tastes, sensations, sights, and sounds? They can only be comprehended by smelling, tasting, feeling, seeing, and hearing. Just so with the peculiar calm, soothing, restful, light feeling that is the most frequent result of massage, which cannot be understood until experienced. It doubtless arises to a great extent from the pressure of natural worn-out *débris* being speedily removed from off terminal nerve-filaments. Furthermore, massage excites and awakens the *muscular sense* in an agreeable and beneficial manner such as nothing else does, restoring idio-muscular contractility and extensibility ; and we know that the state of our muscles indicates and often determines our feelings of health and vigor or of weariness and feebleness. Estradère sums up the effects of massage in the following beautiful language: "I think that this happiness, this quietude, this respiration more free, these ideas so pleasing, are the result of the equilibrium which at this time reigns in all the functions. The nervous system, no longer requiring to exert itself against obstacles to respiration, to circulation, and to nutrition, enjoys a tranquility almost equivalent to repose, and then this state of oblivion *de la vie expectative*, in some manner leaves the imagination to dwell upon the ideas of beatitude which come in multitudes to occupy the nervous centres, and these now have no need to concentrate a certain part of their activity to control the functions — to subdue some and to stimulate others."

CHAPTER V.

"The veins unfilled, our blood is cold and then
We pout upon the morning, are unapt
To give or to forgive; but when we have stuff'd
These pipes and their conveyances of our blood
With wine and feeding, we have suppler souls."
— *Coriolanus*, Act V., Scene i.

THE late Professor Biddle, of Jefferson Medical College, than whom a more stately orator never graced a lecture-room, used to wind up his course of lectures on materia medica and therapeutics by saying in the most earnest manner, " and, gentlemen, when everything has failed, you must advise your patients to fall back upon the comforts and consolations of religion." Nowadays, when everything fails, it is the fashion to fall back upon massage, and this is too often expected to act like an antidote to poison, and indemnify patients for what they are apt to regard as a vexatious loss of time. If benefit results, they wonder why this was not suggested or tried sooner. In 1872 I wrote an essay in which I ventured to predict that, at the rate at which nervous diseases were on the increase, some means would surely arise to better aid in their treatment, and possibly help in preventing them, and that this means would very likely be massage. Here are the words: "Statistics show that nervous diseases have been keeping pace with civilization at a fearful rate. In the city of Chicago, the average proportion of neural deaths to the total mortality was, in the five years beginning with 1852, 1 in 26.1. In the five years from 1864 to 1868 inclusive, the proportion was 1 nerve death to every 9.9 of all deaths." [1] Such inroads upon human health are

---

[1] Wear and Tear, by Dr. S. Weir Mitchell.

not likely to continue long without some fortunate means spring-
ing up of preventing, alleviating, or curing them ; not the least
amongst these, and perhaps the most natural and agreeable will
probably be massage."[1] Time has justified this prediction, and
every physician who within the past fifteen years has taken any
interest in massage, is amazed at the apathy of his brethren
towards the value of this treatment in properly selected cases,
while the brethren have always had a private opinion of their
own that rubbing was a good thing if it could only be done
in the right manner.   But this interest is not limited to any one
branch of medicine, for concerning the extent of the usefulness of
massage it may with safety be said that, at tolerably definite
stages in one or more classes of affections in every special and
general department of medicine, evidence can be found that it has
proved either directly or indirectly beneficial or led to recovery,
when other means had been but slowly operative or apparently
had failed altogether.   According to the requirements of individual cases, massage may be of primary importance, or of secondary
importance, of no use at all, or even injurious.   In what affections
and at what stages is massage beneficial ? Briefly answered, for the
present, in local and general disturbances of circulation, locomotion, and nutrition in their incipient stages or after the acute
symptoms have passed away.   At the commencement of many
affections, local congestion and irritation will often be relieved
by massage, and this may serve for cure or prevention of further
mischief.   In cases that have come to a stand-still or lapsed
into a chronic condition, languid circulation will be aroused,
waste products absorbed, and nerves and muscles nourished and
strengthened.   So much for general considerations.   Let us be
more special.

In recent times, *as an auxiliary* in the treatment of neurasthenia and impoverishment of the tissues in females, massage has
won its greatest merit under the direction of Dr. S. Weir Mitchell,
who, as artist and architect, has succeeded admirably in combining the very natural remedies of rest, quiet, food, massage, and
electricity, so as to build anew the broken-down constitutions of

---

[1] Graduating Thesis.

many hopeless invalids. These are patients who have got into a state of *hibernation* from some physical or mental strain, less often from sheer indolence, so that they have become so weak and wakeful and full of aches and pains that they cannot get up until their blood has been replenished and their tissues reformed, until they have been soothed and comforted to sleep by the warm hands of the manipulator, and their muscles aroused to action by lightning strokes skillfully let loose upon them. Then at last comes the welcome command, inspired by experience and judgment, to get up and walk. Certainly this is wonderful, considering that it is usually accomplished in the brief space of two or three months against illnesses that have kept patients in bed sometimes for years. Playfair mentions cases that were cured by this combination of treatment, who had been bedridden invalids for six, nine, sixteen, twenty, and twenty-three years respectively, and some of these were such sufferers that they had to be transported from their homes to London for treatment under an anæsthetic. A comparison of Playfair's cases with those of Mitchell would tend to show that cases of nervous prostration and hysteria in the higher walks of life get into a much worse condition in England than they do in America. It is presumed, however, that they are more numerous in America than in England, on account of atmospheric peculiarities, high pressure in education, greater social claims, and less physical exercise. But this is by no means certain, if we may judge by the present alarm of English sanitarians about the evil tendencies of their system of education and cramming, and other matters alike prejudicial to health in England as in America. The best summing up of the nature and symptoms of the cases under consideration is given by Goodell in his "Lessons in Gynecology," as follows : "The general pathology of such a neurosis is not perfectly clear, but it probably consists, as Beard first pointed out, essentially in malnutrition of nerve-centres, followed by disturbances in the circulation from weakened innervation. These secondary disturbances consist of local anæmias and of local hyperæmias. In other words, in that equilibrium of wear and repair which means health, a disturbance occurs, which means disease. There will be sudden ebbs and flows of impoverished blood in the various vital organs — the same kind of surface flushings and blanchings going on in the deeper structures. Thus we may see in the same person, and starting from one cause, alternations of anæmia and hyperæmia of brain, stomach, or spine. The cerebral exhaustion or irritation manifests itself by clavus, wakefulness, heaviness, asthenopia, and by inability to read or to write, or to concentrate the thoughts on any given subject; the exhaustion of the stomach by flatus, nausea, gastralgia, capricious appetite, and so on; the spinal exhaustion by tender spots, backache, and weariness. The anæmia of the reproductive organs is exhibited by amenorrhœa or by scant menstruation, by neuralgic and hysterical pains; the hyperæmia by congestion, dysmenorrhœa, menorrhagia, and leucorrhœa, by uterine flexions and dislocations, and by a variety of subjective and objective phenomena with which every physician is familiar."

After a careful diagnosis, which is by no means always easy, in order to decide that there is no organic disease of the nervous system, the essentials of the treatment consist in secluding the patient from over-sympathetic friends, in absolute rest, in carefully regulating the diet by commencing with small, definite quantities of skimmed milk, and gradually increasing this and other food until incredible quantities can be assimilated, and all this is made not only possible but decidedly advantageous, by the daily use of massage and electricity. It is needless to say that this combination of treatment, in order to be successful, must be dispensed under the guidance of a firm, kind, and skilful physician, who studies each case on its own merits, varies the treatment according to circumstances, and administers tonics and laxatives as they are indicated. Local treatment when necessary does not conflict with the measures first spoken of, though in general the advocates of the rest-and-food treatment prove that there has been overmuch topical treatment. In the therapeutics of these cases, the first place is accorded to massage; but just what is meant in the chapter on Electricity in "Fat and Blood," by the remark that "no such obvious and visible results are seen as we observe after massage" is not very clear, however true. What obvious and visible results? Are not the contraction of the muscles and the elevation of temperature caused by electricity more evident than the less immediately apparent effects of massage? Not any clearer, but equally valuable as the opinion of an acute observer, is the brief sentence of Prof. Playfair, that "electricity forms a valuable subsidiary means of exercising the muscles." The manner of using the electricity is of interest. Slow interruptions of an induction current, one in every two or five seconds are preferred. These allow distinct contractions and relaxations of the muscles, and cause dilatation of the blood-vessels. They are, therefore, closely imitative of the physiological functions, as is massage. The difference between the continuous and interrupted currents is considered by some good observers to be owing to the difference in their time of passing. "The faradic currents are lacking in the chemical power of the continuous current because they pass so quickly that they have not time to exert a chemical influence. Gunpowder can be passed so quickly through the hottest flame as not to ignite it." This leads us to say that massage is often done with too great rapidity, especially for its best effects in nervous cases, for it takes time to communicate sensation, and if sensations are painful, they take four times as long to pass through the spinal cord and reflex action is excited, which means that the patient involuntarily resists the impression. I have elsewhere stated that percussion is to massage what faradization is to electricity, and will often answer the same purpose; manipulation or deep-kneading is to massage what the constant current is to electricity, and the ultimate effects of each are very much alike. In *Schmidt's Jahrbücher* and elsewhere, instances are recorded in which massage has succeeded after electricity and other means had failed. The reverse of this may be true, but as yet the proof of it has not come before me. Massage is seldom disagreeable, even from the first, if sufficient tact is used. When, however, at its commencement it is unpleasant, it soon becomes acceptable, so that it is looked forward to as a luxury and a pleasure, and spots that are sensitive, tender, and painful can gradually be encroached upon and soon disappear. It is not unlikely that as the quality of the massage improves the quantity necessary for these neurasthenics will be less. The fact that grease is so constantly and freely used is an indication of the quality; and two doses of an hour to an hour and a half daily seem inordinately large and frequent. When a patient is at all subject to the influence of massage, subsequent applications of the same force and duration as the first have an increased effect, so that if half-hour applications at first had any apparent effect, the result would probably have been the same, even if the time had not been extended. It is but fair to say that the later massages may be much longer and stronger than the earlier, and indeed there may be no end to the acquired tolerance and pleasure of the patient to the vigor and duration of these treatments. They may end in useless indolence and luxury unless there is a clear head and a strong will to push the patient along, and supply the missing link between will and action. It is still a question to be decided what the minimum of massage may be that will do the greatest good in any given case. Many a time I have relieved a tired headache and put a wakeful patient to sleep by ten minutes' massage of the head alone, when subsequent applications of longer duration, including both head and body, had no greater effect. Who knows but we may yet have a homœopathy in dosage of massage? As to the frequency of applying massage, common sense would say let it be repeated before the effect of the previous treatment had passed away, so that a cumulative effect may be obtained; and as to the length of time the operations may be continued, common sense would again speak and say, so long as the patient improves, or until recovery results. In the absence of accurate doses of massage, Mitchell and Playfair have adopted the safest plan by using probably more than sufficient, and pushing this and all the other means previously mentioned with an energy from which there was no appeal. The results have more than justified the methods, and the best cures have been obtained in the worst cases of long standing, bedridden, and wasted invalids. There is a definite reason for this not yet explained, but which becomes clear

on the application of the principles of treatment here incul-
cated. Those patients who have had a long period of rest, have
already undergone in this a great part of the treatment necessary to make them responsive to the remainder ; while with
others who still struggle on, growing worse at every effort, it is
necessary that they should be put to bed and absolute rest enforced.
From this combined method of treatment, Prof. Playfair says he
has had more satisfactory and surprising results than he has ever
before witnessed in any branch of his professional experience.[1]
Of course it is not likely that any but men of great prestige and
long experience would be intrusted with the care of such apparently hopeless invalids, where the trouble and expense of
carrying on the treatment are so great. A few illustrative cases
will probably be of interest.

On September 10th, a gentleman came to consult Professor
Playfair about his wife. He stated that she was then fifty-five
years of age, and had passed ten years of her married life in
India. At the age of thirty she was much weakened by several
successive miscarriages, and then drifted into confirmed ill-
health. They had been married for thirty-four years, of which
the last twenty had been spent by her in her bed or on the sofa.
She was unable even to stand, and found the pain in her back
too great to admit of her sitting up. She was utterly without
strength, of an intensely nervous temperament, and suffered
incessantly from neuralgia. There was not the slightest symptom
of paralysis. She did not take morphia, nor any narcotic stimulant of any kind, beyond a glass or two of wine daily. That she
had long been in a state of hysteria was the opinion of nearly all
the medical men who saw her. Although the attempt to cure so
aggravated a case as this was certainly a sufficiently severe test
of the treatment, it was determined to make the trial, and the
patient was removed from her own home and isolated in lodgings.
She was found in bed supported everywhere by many small pillows
and wasted more than Professor Playfair had ever seen any human
being. Though naturally not a small woman, her height being
five feet five and a half inches, she only weighed sixty-three

[1] Page 85, Nerve Prostration and Hysteria.

pounds. No organic disease of any kind could be detected.
The appetite was entirely wanting, and she took hardly any food
beyond a little milk and a few mouthfuls of bread. From the
first the patient's improvement was steady and uniform. The
way she put on flesh was marvellous, and one could almost see
her fatten from day to day. Within ten days all her pains, neuralgia, and backache had gone, and have never been heard of
since, and in that short space of time all her little pillows and
other invalid contrivances were got rid of. Her food diary on
the tenth day after the treatment was begun may be of interest;
and all this was not only consumed with a relish, but perfectly
assimilated by this bedridden patient who had lived on starvation
diet for twenty years.

6 A.M., ten ounces of raw meat soup; 7 A.M., cup of black
coffee; 8 A.M., a plate of oatmeal porridge, a gill of cream, a boiled
egg, three slices of bread and butter, and cocoa. 11 A.M., ten
ounces of milk; 2 P.M., half a pound of rump steak, potatoes, cauli-
flower, a savory omelette, and ten ounces of milk; 4 P.M., ten
ounces of milk and three slices of bread and butter; 6 P.M., a cup
of gravy soup; 8 P.M., a fried sole, three large slices of roast mut-
ton, French beans, potatoes, stewed fruits and cream, and ten
ounces of milk; 11 P.M., ten ounces of raw soup.

The same scale of diet was continued during the whole treatment, and never produced the slightest dyspeptic symptoms. At
the end of six weeks from the first day of treatment she weighed
one hundred and six pounds — a gain of forty-three pounds. In
eight weeks from the commencement of the treatment she was
dressed, sitting up to meals, able to walk up and down stairs with
an arm and a stick, and had also walked in the same way to the
park. This was more than had been hoped for, and soon after she
left with her nurse for Natal, and no doubt she would return from
her travels with her cure perfected.

Another case of Professor Playfair's was that of a young lady
suffering from intense hysterical vomiting which had commenced
six years previously, after severe mental strain. Latterly she
could keep nothing on her stomach but a single mouthful of
milk, and this only when mixed with whiskey, so that in this

way she was taking three or four glasses of spirit daily.   She was terribly emaciated, weighing only sixty-three pounds.   Her mother wrote of her, " It is just five years last Christmas day since she has retained a single meal.   Her symptoms have been most distressing and have resisted every kind of treatment. Her young life has been completely blighted, and I have long since given up her case as hopeless."   The rapidity of the cure, in this instance, was almost ludicrous.   In three days after she was isolated she was keeping down two quarts of milk, and this no longer with the aid of whiskey.   In ten days she was eating with an enormous appetite, and in six weeks she left weighing one-hundred and six pounds, a gain of forty-three pounds, and has since remained quite well.

Illustrations of the success of this method of treatment are given by Dr. Mitchell as follows : " Miss C., aged twenty-six, at the age of twenty passed through a grave trial in the shape of nursing her mother through a typhoid fever.   Soon after a series of calamities deprived her of fortune, and she became for support a clerk, and did for two years eight hours of work daily.   Under these successive strains her naturally sturdy health gave way.   First came the pain in the back, then growing paleness, loss of flesh, and unending sense of tire.   Her work, which was a necessity, was of course kept up, steadily at first, but was soon interrupted by increase of the menstrual flow with unusual pain and persisting ovarian tenderness.   Very soon she began to drop her work for a day at a time.   Then came an increasing asthenopia with evening headaches, until her temper changed and became capricious and irritable.   When I saw her she had been forced to abandon all labor, and had been treated by an accomplished gynæcologist, and was said to be cured of prolapsus uteri and of extensive ulceration, despite which relief she gained nothing in vigor, endurance, and got back neither flesh nor color.

" She went to bed December 10th, and rose for the first time February 4th, having gained twenty-nine pounds.   She went to bed pale and got up actually ruddy.   In a month she returned to her work again, and has remained ever since in health, which enables her, as she writes me, ' to enjoy work and do with myself what I like.' "

" Two years ago, Miss L. came to me with the following history : At the age of twenty she had a fall, and began in a week or two to have an irritable spine.   Then after a few months a physician advised rest, to which she took only too kindly, and in a year from the time of her accident she was rarely out of bed.   Surrounded by highly sympathetic relatives, to whom chronic illness was somewhat novel, she speedily developed with their tender aid hyperæsthetic states of the eye and ear, so that her nurses crept about in a darkened room, the piano was silenced, and the children kept quiet.   By slow degrees a whole household passed under the selfish despotism of a hysterical girl.   Intense constipation, anorexia, and alternate states of dysuria, anuria, and polyuria followed, and before long her sister began to fail in health, owing to the incessant exactions to which she too willingly yielded.   This alarmed a brother, who insisted on a change of treatment, and after some months she was brought on a couch to this city.

" At the time I first saw her, she took thirty grains of chloral every night, and three hypodermic injections of one-half grain of morphia daily.   As to food she took next to none, and I could only guess her weight at ninety pounds.   She was in height five feet two and a half inches, and very sallow, with pale lips and the large indented tongue of anæmia.   I made the most careful search for signs of organic mischief, and finding none, I began my treatment as usual with milk, and added massage and electricity without waiting.   Her digestion seemed so good that I gave the iron in twenty-grain doses from the third day, and also the aloe-extract pill thrice a day.   It is perhaps needless to say that I isolated her with a nurse she had never before seen, and that for seven weeks she saw no one else save myself and the attendants.   The full schedule of diet was reached at the end of a fortnight, and the chloral and morphia were given up at the second day.   She slept well the fourth night, and save twice a slight return of polyuria, she went without a single drawback.   In two months she was afoot and weighed 121 pounds.   Her change in tint, flesh, and expression were so remarkable that the process of repair might well have been called a renewal of life.   She went home, changed no less morally than physically, and resumed her place in the family circle and in social life, a healthy and well-cured woman."

There has been and still is a muttered growl of discontent on the part of many of the profession that the author of "Fat and Blood" has only published a few of his successful cases. No one can accuse him of lack of candor if they will bear the following words of his in mind: "I might multiply these histories almost endlessly. In some cases, I have cured without fattening; in others, though rarely, the mental habit formed through years of illness has been too deeply ingrained for change, and I have seen the patient get up fat and well only to relapse on some slight occasion."

The author of "Nerve Prostration and Hysteria" sums up his experience in saying that, on looking back at his cases, the only ones with which he had any reason to be disappointed were those in which the primary selection had been bad; and in the few in which the results were not thoroughly satisfactory, he had doubt as to their suitability for this treatment which he expressed before-hand. These included one case of chronic ovarian disease, and one of bad anteflexion with fibroid enlargement of the uterus, in both of which the local disease prevented really beneficial results. In a third case, he had to stop the treatment in a week on account of cardiac mischief; two others were cases of positive mental disease; and in one there was epilepsy. Whether it was the rest, seclusion, and feeding, or the massage and electricity that seemed to be at fault in the case of cardiac trouble is not stated. In one case of nervous prostration with old-standing uterine disease in which a fibroid tumor had been removed by Spencer Wells, but which sub-sequently grew again to nearly the same size as before, and was then accompanied with endometritis, vaginitis, piles and prolapsus of the rectum; in which nutrition was at the lowest ebb and the patient taking large doses of morphia and chloral, Prof. Playfair undertook the treatment reluctantly, but his efforts were crowned with ultimate success. During the last month of treatment, the uterine symptoms were not complained of and were not inquired about. Of course, the uterine fibroid could not be removed; but many women have similar growths that do not affect them.

Goodell mentions the following cases to show that the success of the treatment here referred to does not always depend upon a rapid increase of flesh: "Miss K. R., who had excruciating suffering at her monthly periods, defective locomotion, and other marked uterine symptoms, besides great nervous prostration, became well, although she gained but five pounds. Mrs. M., a sterile lady with a heavy and tender retroflexed uterus, and with prolapsed ovaries, was wholly relieved of ovaralgia, menorrhagia, and other grievous symptoms which for years had embittered her existence, yet her gain was but seven pounds. Miss P., also with prolapsed ovaries, and with a coccygodynia so severe that at one time serious thoughts were entertained of removing the coccyx, was restored to health with but little gain in flesh." These cases got well with little or no local treatment; the essential part of the disturbance, in Dr. Goodell's opinion, being in the nerve-centres, and requiring only rest and nourishment, sleep and freedom from pain, and as means of obtaining these, massage and electricity.

By far the most elegantly expressed paper on the rest and feeding treatment is from the pen of Dr. F. W. Page, whose extensive experience as physician at the Adams Nervine Asylum has favored him with advantages for its study and observation enjoyed by few, if any. The article is entitled "The Permanency of the Rest Treatment," and is published in the *Boston Medical and Surgical Journal*, 1882, page 77. The author tells us that in order to overcome the evil effects of prolonged rest, massage is employed, and the next element, which is merely an auxiliary, is electricity, the effect of which is to accomplish what can usually be done much better by manipulation, and with less pain and discomfort. When applied with a weak current and slow interruptions, he finds it usually agreeable. Proofs of cure are founded on personal examinations after one or two years' time. The length of time during which the majority of these patients had been invalids varied from two to fifteen years, and the average duration of the treatment of patients in the Asylum was three months and three weeks. The patients who recovered made an average stay of four months and three weeks, while those not relieved only made an average stay of two months and twelve days. In a later report, the following observations of Dr. Page are full of interest and instruction: "The value of rest and seclusion as remedial agents is an exceedingly interesting question, in view of the rather indiscriminate application which has been made of this treatment to all forms of nervous disturbances, whether of cerebral, spinal, or peripheral origin. This method of treatment is of great value in many disorders. In hysteria it assists by its discipline in regaining self-control, but *not* in melancholia. On the contrary, its use in cases of *depression* invariably aggravates rather than soothes or mitigates the symptoms, and I do not now resort to it in the treatment of this class of patients. In nervous exhaustion, so-called, whether of cerebral, spinal, or mixed types, it is more certain of satisfactory results than any other plan yet tried. It retards waste, checks morbid activities, and prevents the drains and lesser strains so fruitful in perpetuating this condition. Whether imaginary or real, the confused thought; the supposed loss of memory or its sudden lapses; the weakness of vision; the irregular and varied forms of circulatory disturbance; the disturbed digestion; the motor irregularities; incapacity for work; insomnia and all other symptoms speedily disappear under its judicious use, seldom to reappear. Neurotic traits, whether psychological or not, manifest themselves in parents and children for successive generations. Even disturbed nerve-functions, like migraine and neuralgia, perpetuate themselves in different members of a family through a long line of descent. Always quiescent in good health, these neuroses promptly manifest themselves whenever nutrition is seriously disturbed."

Before the publication of the valuable experience of the gentlemen above mentioned, Dr. Henry B. Stoddard and the writer published the results of massage [1] alone in a few cases similar to those here referred to. Our use of this treatment in such cases was unknown to each other until Dr. Stoddard sent me a report of his. My first case is of interest as showing that when everything had apparently failed, and exercise, though it could be taken, was of no use, yet massage improved the physical condition of the patient without corresponding gain in the mental. In this last respect, the truth of the conclusions of Playfair and Page is striking; not, however, to the extent of regarding massage as contra-

---

[1] Massage in Amenorrhœa and Dysmenorrhœa. Boston Medical and Surgical Journal, February 10th, 1876.

indicated, and likely to aggravate depression or mental disease ; for I take it that they have reference rather to rest and seclusion, and to massage only as a means of employing these without impairment of nutrition. Miss F., twenty years of age, was a short, stout lady, well nourished, slightly adipose, of a plethoric, ruddy appearance, and good muscular vigor. She had never studied too hard, nor overworked in any way, but had always seemed to be in a depressed state of mind. She complained at times of fulness in the head with slight vertigo and general discomfort. At fourteen years of age, she began to menstruate, and this function continued to recur for three and a half years, but was always from ten days to three weeks too late. The menses then ceased for two years and a half, and the best experts could discover no local cause to account for the interruption. During this interval, tonics, emmenagogues, horseback riding, and a year's travel in Europe were tried, but without effect. Though the physical condition of the patient seemed good, her body well nourished, her complexion ruddy, her sleep and appetite fair, yet she became more and more low-spirited, for which she, or rather her relatives, sought the advice of Dr. John F. Tyler. He advised a course of massage. In January, 1875, when I began giving her massage, she had been in the habit of walking two or three miles daily ; but it was only from a sense of duty and with great effort and disinclination that she did anything, physical or mental. There were no sanguine expectations on her part of benefit from any treatment whatsoever.

The mode of procedure was manipulation of the limbs, back, and abdomen, with percussion to the back and resistive movements of the feet, legs, and thighs, in all their natural directions, together with passive and resistive circumgyration of the trunk. The treatment was administered for three-quarters of an hour, every other day. After seven such treatments, the menses came and lasted three days, though rather scanty, having been suspended for two years and a half. Throughout the following month, the same treatment was continued, with a view of increasing nerve-force and correcting probable vaso-motor disturbance, but resistive movements were omitted until within a week of the next expected monthly period. The menses, this month, appeared five days later than what was considered the proper period, though formerly at the best they had always been from ten days to three weeks too late. Treatment was now discontinued until within a week of the next expected return. The catamenia this time did not appear until nineteen days after the proper date; and this delay was attributed by the patient's highly intelligent relatives to the discontinuance of massage for three weeks. This view would seem to be favored by the result of the ensuing month, for massage was resumed two weeks prior to the hoped-for event, and exactly twenty-eight days from the last period the catamenia again returned. The next month the patient had four visits in the ten days preceding the natural time for recurrence, and this took place one day earlier than the normal interval. The quantity each time after the first was what the patient considered to be about natural. Thus far the treatment seemed to have an increased effect and the patient an increased susceptibility to it, even though it was lessened in frequency and not increased in quantity. Without further massage or other treatment, all the functions continued regular for several months, and a year and a half passed before I again heard from the patient, when I was sent for and learned that the catamenia had been absent for three months in spite of all ordinary remedies. Massage was again resumed with even more favorable results than before, and in a few months the patient started for Europe. At the time of her departure, her mental condition did not seem to have improved, but I have since learned that it did afterwards to a marked degree, and that the beginning of this, partial and impartial observers agree in dating from the time massage began.

At the first visit, this patient's tissues seemed to me to be exceedingly dense, matted, and inelastic; at the fourth visit they were much more supple and elastic, and continued so to be, in consistence with her easy mode of living.

In marked contrast with the condition of the tissues in the previous case, and the effect of massage upon it, is the flabby, atonic state of muscles resulting from long illness in the following case, which might well illustrate that "hard rubbing binds," while the one just related might show that "soft rubbing loosens." After unusual exertion and anxiety in nursing her mother and sister, Miss A. suffered great nervous prostration. The trouble at first, the patient said, was all in her head; she was very wakeful and had frequent attacks of hysteria. Several months later she was seized with intestinal catarrh, and as this was accompanied with great pain, it aided very much in reducing her. From this she gradually recovered so as to be able to sit up for a few minutes at a time. A persistent backache and profuse leucorrhœa called attention to the uterus, which was found to be anteverted. When tenderness had subsided so as to admit of a Hodge's pessary, this, with a bandage around the abdomen, afforded great relief. Menstruation was regular as to time, though painful and scanty, lasting but a day, and passing only when the patient was sitting up. In the mean time, the hysteria continued, at times closely simulating peritonitis, and her physician, Dr. J. T. G. Nichols, of Cambridge, from whom I got the main points of her history, informed me that hysterical convulsions, mania lasting from a few hours to several days, and transient aphonia were also of common occurrence. Injections of assafœtida alleviated these attacks, and a course of tonics and electricity had improved her so that she could be up four hours daily, an hour or two at a time. Excepting the occasional use of a vegetable bitter, nothing had been administered for two or three months, when Dr. Nichols kindly referred the patient to me for massage, because he thought that the muscles might in this way receive the exercise which they so much needed.

Massage was begun in May, 1875, when the patient had been an invalid for over two years. At this time she was taking nothing but a gentle laxative every day. For a while at first careful manipulation alone had to be used in this case, as anything like passive or resistive motion, except of the feet and arms, was very apt to give rise to abdominal pain, which was frequently referred to one or the other of the ovarian regions, and sometimes followed by hysterical convulsions. After my second visit, the laxative was laid aside in the hope that the kneading of the abdomen would produce the same effect; and in this we were not disappointed, as she had a natural daily dejection without medicine. She had massage two or three times weekly for ten times before her next monthly period. When this arrived the menses came, somewhat to our surprise, while the patient was lying down. This was the first time that the flow appeared in that position for sixteen months. She was under massage for two months longer, and in each the menses came with lessened discomfort, and while the patient was in the recumbent posture. With regard to the aches and pains, those of the back and head, as well as the uncomfortable feelings in the abdomen, were alleviated at each application of massage, and the patient was greatly soothed, sometimes to sleep. The cold hands and feet were made warmer, not merely for the time, but permanently. The muscles gained in size and firmness, and the patient walked with much less scuffling of the feet, and went up and down stairs naturally. But still she was very much of an invalid, unable to ride in a carriage without suffering pain in the back and abdomen, though she could walk short distances with ease. In her great desire to get well, she overestimated her improvement, while the previous case probably underestimated her ameliorated condition. In this second case, the combined treatment of Weir Mitchell would no doubt have brought about recovery.

Dr. Stoddard very kindly sent me his notes of the two following cases in which he employed massage, not by relegating it to the nurse or one of the patient's relatives, as is usually the way, when it is almost sure to be done in a slipshod manner, or what is worse, overdone, but by applying it himself. "A patient of a spare habit and decided nervous temperament, suffering from chronic inflammation of the uterus with ulceration, and also from nervous prostration, in addition to other troublesome symptoms, was the subject of obstinate sleeplessness which had lasted for some time before she came under my care. Nature's sweet restorer was usually sought in vain, till three or four o'clock in the morning, when a couple of hours of uneasy slumber were obtained. At my suggestion, she had used various remedial agents, chloral, bromides, lupulin, hyoscyamus, and morphia, alone or in combination, for several weeks, but with indifferent success. One evening I was induced to employ massage for the first time in her case, in the hope of relieving the reflex pain in the back and limbs, which at that time was especially troublesome and persistent. The application proved so agreeable to my patient, and so promptly and effectually relieved the pain for which it was employed, that at the same sitting I extended its use to the rest of the body, and with a very gratifying result; for soon after I left her, a drowsiness which was then quite evident passed into a quiet and refreshing sleep, from which she did not awake until six A. M. Its subsequent employment in this case never failed, except when severe pain was present, to secure for her a good night's rest. Furthermore, its regular employment three times a week, for three weeks after the first trial, not only much improved the capillary circulation, which had been quite languid previously, but seemed to be largely instrumental in securing a regular and healthy menstrual flow after an absence of at least six months. A variety of emmenagogues had previously been prescribed without effect. I take this occasion to state that, at the time referred to, this patient had been taking for some weeks a preparation of quinine, strychnia, and phosphoric acid, and was under local treatment for the uterine inflammation, and that in my judgment important indications were met by massage when other remedial agents proved inadequate or were but slowly operative. In this case a nervous headache to which she had long been subject was always much alleviated by the application of massage to the head."

Dr. Stoddard's second case is as follows: " I have under my care another patient with uterine inflammation of eight or ten years' standing, characterized by an enlarged and indurated cervix and retroflexed body of the uterus, with marked menstrual irregularity and severe dysmenorrhœa. When she came under my care, some two years since, the approach of the menses was accompanied by excessive pain and a series of hysterico-nervous convulsions. She was also the subject of a variety of reflex symptoms, among which may be cited, as most prominent, pain and tenderness of the sacro-iliac region and over the entire spinal column; sleeplessness; attacks of numbness in the extremities of a very decided character, and attended with flexor spasm; a feeling of pressure at the vertex; dyspepsia, and meteorism. For the dysmenorrhœa and resulting convulsive attacks, the subcutaneous injections of morphia proved the only effective remedy. Happily the local and general treatment had very much modified the tendency to such seizures. But for the relief of some of the more constant reflex symptoms, massage proved a very hopeful agent. Spinal and sacro-iliac pain and tenderness have been very much relieved by its local use, and its regular employment over the whole body three times a week, while not directly inducing sleep, as in the previous case, has seemed to tranquilize the nervous system and render it more susceptible to chloral and other hypnotics; and the attacks of numbness and flexor spasm have been much diminished in frequency and severity during its use. Applied to the head, massage has had a decided influence in temporarily relieving that sense of fulness at the vertex which is so common and annoying a symptom in uterine disorders. The meteorism in this case had been for several years a persistent and troublesome symptom. I have often seen the abdomen distended to as great a degree as if she were at the close of gestation, tense and tympanitic, and productive of marked dyspnœa and cardiac spasm by the upward pressure. Massage, locally applied, has been more effective in relieving this condition than any other means employed. Repeatedly has its thorough application, extended over a period of fifteen or twenty minutes, been followed by a subsidence of the tympanites and a restoration in good degree of the natural softness of the abdomen, with a corresponding relief of pain. Incidentally, the constipated condition naturally attending such atony of the muscular coat of the intestines has been in a measure corrected by the repeated applications.

These four cases are fair examples of what may be expected from massage in doubtful, half-and-half cases which every physician knows are so difficult to manage, and Prof. Playfair says of such that they are the only kind that have caused him disappointment. Dr. Stoddard remarks that many times he would have been at his wits' ends without massage. His observations

of the increased effect of tonics and sedatives, which prior to massage had been comparatively inert, are worthy of notice; for the
more perfect the circulation, whether this be caused by massage
or otherwise, the better will medicines of all kinds be distributed through the system, and the greater will be their effect.
But the inherent qualities of massage alone are also soothing
and invigorating, and often as apparent as in Dr. Stoddard's cases
when all previous medication has been left off.

So often have I observed an increase in the quantity of the
catamenia and an earlier appearance than usual in ladies who
are to all intents and purposes well, and who have had massage of
the back or general massage for some slight ailments, that I have
come to regard this as one of the physiological effects of massage.
Even massage of a leg for a joint or muscular affection is frequently followed by an earlier appearance and a longer stay of
the monthly visitor. In amenorrhœa or dysmenorrhœa, where
neither local treatment nor operative procedure is indicated, massage would seem to be a good means to employ, especially in atony
of the nervous system and when there is not present any abnormal state of the blood nor any of the pelvic organs.

In the cases mentioned in this chapter, it is not considered a
good sign if the limbs persist in growing cold under the stimulation of massage. This means augmentation of the vaso-constrictor
function of the nervous system, while the favorable effects of
massage are manifested by warmth and comfort, and agree with
what we understand to be an increase in the vaso-dilator function.
Even so with vigorous exercise, it is not beneficial if a person
turns pale and his blood-vessels constrict, for too much blood will
be repelled to internal organs. The primary agreeable sensations
of a moderate dose of alcohol are due in a great part to the same
effects as usually result from massage or exercise, increased peripheral circulation and increased capacity of the vessels to accom-
modate it.

How much more massage and general measures of treatment
may yet encroach on the domain of gynæcology remains to be
seen. This will best be determined by the co-operation of the
far-seeing neurologist and the skilful gynæcologist. " Every

one," says Dr. T. Gaillard Thomas, "who has had experience in the treatment of these disorders must have been struck with surprise at the wonderful improvement exerted upon cases which have long resisted local means, by a sea-voyage, a visit to a watering place, or a few months passed in the country. Not only is the improvement manifest in the general state of the patient; it shows itself locally also, and in some cases complete recovery may be obtained." When these fail, try massage. The original compounder of rest, seclusion, and excessive feeding, made available by massage and electricity, deserves the everlasting gratitude of the profession and the public, for in this and other ways he has fulfilled the prophecy of Dr. John Hilton, who said, " As the most precious treasure is the most securely hidden, as the solicitous and patient explorer sees no charm in trophies won with ease, so I am assured that the industrious laborer who will pry with bright and peering vision into the mazes of the nervous system, and apply to the treatment of its manifold derangements the principles of rest, will reap his reward."

It is generally conceded that it is difficult to ameliorate the condition of fat, anæmic, and neurasthenic patients. The results of massage alone in the following case far exceeded our expectations, changing a slow and tedious recovery into rapid and daily improvement, and leaving the patient in a better condition than she had been for years. Mrs. K. was about thirty-two years of age when I was called to see her on the 12th of June, 1877. Eight years before this she had suffered from a sun-stroke from which she thought she had recovered in a week, but at the end of four weeks severe headaches, nausea, and vomiting would be caused by any slight annoyance, and these would last for several hours, in consequence of which she was obliged to give up a responsible position which she had hitherto filled with energy and ability. Her left side, arm, and leg also troubled her from this time with numbness and slight difficulty of motion apparent to others. A vacation of seven months was of great benefit to her, but she only partially recovered. From the time her health became impaired by the sun-stroke, she grew very fat, soon weighing two hundred and ten pounds, seventy pounds more than her usual weight, and this continued. At times she had been melancholic, and only strong religious convictions had kept her from committing suicide. During these eight years she had suffered at one time or another from ulceration of the cervix uteri, scanty and difficult menstruation at long intervals; piles, fissure of the anus, and prolapse of the rectum; pseudopia, photophobia, purulent inflammation of the ear, etc., etc. Thirteen weeks prior to my first visit, she had given birth to a still-born child which she had carried several weeks past the full time. After confinement she was said to have had milk-legs. When first seen by me, she could only take a few steps by leaning on two chairs, and then the legs, already somewhat œdematous, swelled still more and became livid. There were pain and tenderness over the whole of both legs, but especially in the course of the internal saphenous veins, and on the posterior aspect of the limbs. It had only been for two weeks that she was able to step at all, and four weeks altogether that slight improvement was apparent. For four or five years she had been subject to what she called "numb spells," which would come upon the receipt of sudden news, and they had continued after confinement. At such times she was conscious of her hands, arms, and legs becoming rigid and contracted, which she would try to resist, but soon she would become unconscious and remain so for about an hour. When coming out of these attacks, she would be conscious for several minutes before she could speak. Wakeful nights, morning headaches, weary days, and poor appetite were constant since confinement. In addition to all these troubles, the patient was intensely hyperæsthetic all over, so that the slightest touch would cause involuntary recoil. The joints were exquisitely sensitive and rebellious to passive motion. Well, then, how was massage accomplished? Certainly not by friction, but by applying the hands slowly, gently, and firmly with deep manipulation. Thirty minutes of massage was thus given at the first visit to the feet, legs, and outer aspects of the thighs, and a few minutes to the head. During this the sensitiveness decreased, and the manipulations, from the first approach of which she flinched, became agreeable. The patient felt a prickling sensation from the pressure of the hands, which she thought felt just like the sponges of a galvanic battery. The morning headache was relieved, and warmth and comfort prevailed in the legs which hitherto had been cold. At the second massage on the following morning, stronger manipulation was borne, and the prickling feeling was less. At this visit, massage was extended over a greater surface, but the back and abdomen were still too sensitive to be touched. The morning headache was again dispelled, and the patient made cemfortable for the day. At the hird visit, general massage was well tolerated, including back and abdomen. Relief of headache was immediate, and comfort and sleep followed. On the evening of this day she walked naturally, not sliding her feet along as formerly. The time for the massage was then changed to the evening, and good sleep with increasing appetite followed, and the patient made rapid improvement from day to day, in place of slight improvement only apparent from week to week, as before massage began. At the end of seven days she went over two flights of stairs to her meals; at the end of nine days, she was assisting in her housework; on the thirteenth day, walked out, and by this time it was evident by the slackness of her dress that she was losing fat. Twenty-six visits were made in thirty-nine days, and the patient made a most excellent recovery; tonicity, firmness, and elasticity of tissues took the place of the soft, loose hydræmic condition; the constipation, piles, and prolapse of the rectum disappeared not to be heard of again; the menses made their appearance but five days overdue and three times the usual quantity (for years they had been nine or ten weeks behind time, and very scanty and preceded by great depression of mind). At this period there were scarcely any premonitory symptoms. It ought to have been said that the lochia pursued a natural course, and the catamenia appeared ten weeks after confinement. They recurred at nearly normal intervals, with the exception of two suspensions for very good reasons, and she is now the mother of two fine children. The "numb spells" and unconsciousness did not return. Rochelle salt was the only medicine used for a few days at first, to soften the hard fæcal masses. Her diet was left to her natural inclination.

It might seem strange that massage should be tolerated at all in such a hyperæsthetic individual. Not more strange than the fact that in strychnia poisoning convulsions are excited by the slightest touch, a draught of cold air, or a loud noise, whereas firm grasping or deep kneading of the muscles is frequently grateful. The prolapse of the rectum was severely handled in being replaced after labor, and it was agreed that the piles must be operated upon as soon as the patient's strength would permit. That this has not been necessary was an agreeable surprise. It is a matter of frequent observation that those who lift and strain at heavy weights are very apt to suffer from hemorrhoids. This may be accounted for by the forcible and long-continued con- traction of the muscles driving the blood to internal organs where there is less resistance.[1] Massage induces just the oppo- site condition and aids materially by its pressure in pushing along the blood in the portal vein and its branches, "which are capable of holding the total amount of blood in the whole body" (Foster). At the same time, massage stimulates the peristaltic action of the large intestine, relieving the constipation which often causes and increases this trouble, and a free circulation is created in the skin and muscles which should have in them about one-fourth of the total amount of blood in the body, and thus the equilibrium of the circulation is restored. In the *Clinic* for March, 1878, Dr. S. W. Garrod states that he has never failed in treating successfully hemorrhoids of the size of a cherry-stone by punching and kneading them between the thumb and fore- finger and thus obliterating them. It is often necessary to re- peat this operation several times. It is a method with which I have had no experience, but it does not remove the cause of the disease. No one will be so foolish as to suppose that either local or general massage will take the place of hygienic medical and

---

[1] Sedentary occupations, high living, and too little exercise have the same effect, and may also be counteracted to some extent by massage. Too much or too little exer- cise of organs alike enfeeble them.

surgical treatment of diseases of the rectum. But in justice to myself I wish to say here that this patient was so strongly convinced of the efficacy of massage for the relief or cure of hemorrhoids that nothing would do but a friend of hers who was said to be suffering from hemorrhoids must try massage also, and so she applied to her physician for assent, not for permission, for she had previously made up her mind. From a tedious stay in bed of seven weeks after confinement, this patient rapidly gained flesh and strength, but with no lasting relief to the rectal distress. When she was operated on, in place of hemorrhoids, there were found several small solid tumors which were removed. The patient appreciated her physician's and surgeon's derision of massage by consoling herself that inasmuch as it had put her in excellent condition for the operation, it might also enable her to get over it faster, and it did. But for something more than hemorrhoids in this case, the honor and glory of an operation might have been saved. Massage is not infrequently used for preparing a patient for an operation and aiding their recovery afterwards.

Dr. Gage, of Worcester, kindly referred to me for treatment a lady who is at present under my care. From hurry, worry, and irregular action of the heart she broke down in the High School ten years ago. She is thin and anæmic, but not at all nervous. Previous to her coming to me she slept steadily, but not refreshingly, her appetite was indifferent, she had no ambition to do anything, was tired all the time, and could only walk with difficulty about a mile. The patient had been losing ground for two years. There had been much sickness and death in the family, and she had been nurse. Menses have always been accompanied with pain, except at last period, when she was weak and prostrated all over. With three-quarters of an hour's massage daily, and a milk diet to begin with, she was in two weeks eating three full meals and two lunches every day, besides taking a teaspoonful of dialyzed iron and a tablespoonful of Trommer's extract of malt at each meal, and all with a relish. Weakness of the limbs had disappeared, and she could outwalk her well sister. Sleep was refreshing from the first massage, and the patient continues vigorous and buoyant, and it is not yet the end of the third week of treatment. At first only one hour's rest twice daily was advised, and half the usual walking. Now she does not wish to rest at all during the day. The muscles, at first rigid like whip-cords, are rapidly becoming full, elastic, and firm. Anæmic murmurs are disappearing, and the patient goes up two or three flights of stairs with ease. This patient has also repeatedly remarked that the pressure of the hands in manipulation felt like the electrodes of a galvanic battery, but this sensation gradually disappears with improvement.

No greasy substances were used in any of Dr. Stoddard's cases or my own, and it would be well to bear in mind that the very meaning of anointing or lubrication is the prevention of friction, call it massage if you please. In my opinion, it is a positive hindrance to massage on account of the slipping preventing the hands and fingers from seizing, grasping, and kneading the tissues. With suitable diet, superficial friction and anointing favor the nutrition of the external tissues, and the increase of fat in the superficial fascia and deeper structures of the skin, while manipulation or kneading acts more on the deeper structures and muscles ; and whether fat or muscle be increased will depend on the predominance of one or the other.

When I have less favorable experience with massage in cases similar to those here related, I will report it. Except in cases of organic change of structure, or the long-continued force of habit, massage will seldom fail to prove beneficial, unless it is evident that the case is of such a nature as to decidedly contraindicate this treatment.

---

[1] At the next period, the catamenia came naturally without pain, weakness, or prostration.

CHAPTER VI.

MASSAGE OF INTERNAL ORGANS.

MASSAGE OF THE UTERUS AND ITS SURROUNDINGS, WITH A REPORT OF TWO HUNDRED AND THIRTY-NINE CASES.

"Genius breaks from the fetters of criticism ; but its wanderings are sanctioned by its majesty and wisdom."

"MASSAGE and expression being the only resort in the hands of primitive people for the completion of difficult labor, they intuitively, by instinct and by long practice, have brought them to a certain state of perfection, although brute force is more relied upon than dexterous manipulation. The methods are so simple, so natural, and so thoroughly in accordance with sound mechanical principles that they have produced good results. Deprived of the brutality of physical force and aided by science, these very means which have so long and so well served the ignorant, will attain a higher degree of perfection, and will serve by far better the scientific obstetrician."[1] One would think from these words of Dr. George J. Englemann, that educated physicians were not in the habit of making use of such procedures, and that they know but little about them. For many years, — a quarter of a century at least — physicians have been taught and practised massage of the uterus through the medium of the abdominal walls as a reliable means of overcoming inertia during labor, and when there are partial or irregular contractions complicating labor; also when there is post-partum hemorrhage, or danger of the same, and with modified position and pressure to correct malpositions of the fœtus, as well as for the expression

---

[1] The American Journal of Obstetrics, July, 1882.

of the placenta. For fulfilling such important indications as
arousing, equalizing, and strengthening the contractions of this
hollow, involuntary muscular organ under such momentous circumstances, even though applied in a mediate manner, there is
no longer any question of the well-known efficacy of kneading,
squeezing, and pressure. It is a wonder, however, that massage
has not been used earlier for atonic conditions of the uterus other
than those connected with the parturient state, and it is equally
surprising that it has not been more approved and used for the
invigorating of voluntary muscles and nerves which are much
more accessible, and grant a more ready response. If it were as
easily administered as pills, powders, and liquids, its use would
be much more extensive.

  Massage of the pelvic organs should be intrusted to those alone
who have " clean hands and a pure heart," and such a thorough
knowledge of the pathology and treatment of uterine affections
as is possessed by the most accomplished gynecologists. It
should not be confided to any professional manipulator, however
skilful. Credit is given to Major Thure Brandt, a Swedish gym-
nast, for having been the first to use massage in the local
treatment of uterine affections sixteen years ago, in 1874.[1]
Brandt's method excited much adverse criticism which, however,
has passed away as the excellent results obtained became better
known. In the hands of a layman it was doubtless at first used
without proper discrimination, and extravagant results were
claimed. The next to interest himself in this was a physician,
Dr. Gustaf Norström, of Stockholm, who used the treatment
rationally and in cases that he could understand. He found
massage especially successful in chronic metritis that had not arrived at the period of induration ; and after this in the affection
known as hemorrhagic endometritis. He also obtained good results in prolapse of the vagina, and in chronic inflammations of
the ovary. The catamenia, acute and subacute affections, and
pregnancy are contra-indications. In his report of 1876 is given

---

[1] But in 1865, Dr. A. D. Sinclair, of Boston, began using massage of the uterus and its
surroundings with the patient in the genu-pectoral position for the correction of retrover-
sions and retroflections. Dr. Sinclair did not publish his experiences with this method.

his experience which had then extended over two years and a half, and which shows that in 138 cases of chronic metritis he obtained 43 complete cures, and more than 70 nearly complete. Nine cases of hemorrhagic metritis were cured, and in 7 cases of sterility complicating chronic metritis there occurred conception in two soon after the cure had been affected. In the course of his operations he has never had a fatal termination nor the supervention of general peritonitis.

The operation consists in introducing an index finger into the cul-de-sac behind the cervix in such a manner that the posterior surface of the uterus is reached. This is then raised as far as possible, while the fingers of the other hand grasp and knead the uterus through the abdominal walls. Sometimes the uterus is pressed against the walls of the pelvis laterally, or against the posterior surface of the symphysis pubis. Massage acts in these cases by removing and preventing the inflammatory stasis, by producing resorption of leucocytes and elements which have migrated into the surrounding tissues, and by restoring tonicity to the tissues.

In similar cases Dr. Asp[1] has obtained favorable results by means of general massage and medical gymnastics, without any local manipulation. At the time of his report in 1877, he had treated in this way 72 cases suffering from affections of the uterus. Of these, 35 or 48.6 per cent were cases of chronic inflammation of the uterus. Three-fifths of these were married and two-fifths single, 15 recovered, 13 were much improved, and 7 remained as before treatment. The average length of time of treatment was for those who recovered, 8.6 weeks for single women; and 15.4 weeks for the married. Six of the cured cases continued well, three had no relapse at the end of a year, three relapsed at the end of two, nine, and ten months, respectively, without ascertainable cause, and in three others there was relapse after birth or miscarriage. Eleven cases of anteflexion and one of retroflexion were also treated in this general manner.

---

[1] Dr. Asp is director of an institution at Helsingfors for the treatment of diseases in which massage and movements are appropriate. He is also professor of microscopy and pathological anatomy in the University.

Ten of these were in single women. During the treatment the subjective symptoms disappeared and the patients felt perfectly well, the flexions remained unchanged. The average length of time of treatment was 7.6 weeks. Their subjective symptoms, in the opinion of Asp, proceed from a hindrance to the circulation in the uterus which can be removed by means of massage. Four of his cases were chronic inflammation of the surroundings of the uterus (perimetritis, parametritis), and, of these, three improved much in from four to twelve weeks. In one case of myoma and another of fibroma of the walls of the uterus, the general condition of the patients was improved, but of course not the local. Favorable results were also obtained in other affections connected with the uterus, and of the whole 72 cases, 23 were said to have recovered entirely, 34 were improved, and 15 were unchanged. The author concludes by saying that this method merits more attention from physicians in the future than has been accorded to it in the past.[1]

If Robert Burns had been more familiar with the poetry that had been written before his time, he confesses that he would have dared less. In that event the world would have been the loser, and his verses would have lacked their pathos and fervor had he known that the same thoughts had been expressed in other words. Human minds work in the same channel. Dr. A. Reeves Jackson of Chicago wrote to me that if he had known of the published experience of Norström and others, he would have made his paper on " Uterine Massage," which he read before the American Medical Association, and published in the *Boston Medical and Surgical Journal* for Sept. 23, 1880, much more complete. Like Burns and his poetry, it is well he did not, for in that case his article would probably have lacked the clearness, vigor, and attraction of independent thought worked out and well expressed, which caused the article to be quoted by two French medical publications (*Garnier's Annuel* for 1881, and the *Journal de Médecine de Paris*, June 30, 1883), and again re-echoed back to American shores to be quoted from the latter by the

---

[1] Virchow and Hirsch's Jahresbericht, 1878, Vol. II., pp. 570. Norsk Med. Arkiv, Bd. 11., No. 22.

Medical News of Philadelphia of July 21, 1883, as if Dr. Jackson were a Frenchman, and such treatment had never before been heard of on this side of the Atlantic. The French quotations would lead one to infer that the procedure had never before been heard of, on this side of the Atlantic.[1] Dr. Jackson claims no originality, and his paper could not have been more modest.

From the records of 277 gynecological cases, the doctor found 179 or 64.6 per cent denominated subinvolution, hypertrophy, hyperplasia, chronic metritis, and simple enlargement. The great variety of treatment is an indication of the inefficiency of it, Thomas, Scanzoni, Atlee, and others being quoted to show the hopelessness of relief from the usual methods. The author points out that all the causes of uterine enlargement form an obstruction to the return of the venous circulation. The indications are to lessen the undue and partially stagnant supply of blood, to overcome the stasis, and to promote resorption of the excess of the tissue. All the remedies generally employed act by lessening vascular fulness, but massage has proved more efficient in doing this in Dr. Jackson's hands than any other single means. Not every case of uterine enlargement is amenable to squeezing and kneading. In some it might be injurious. Massage is available in the first stage when the uterus is found low down in the pelvis, enlarged, tender, and spongy, and having a doughy elasticity, its sinuses gorged with blood, and newly formed connective tissue in its walls. Displacements and distortions alone do not preclude massage. If the first or hyperæmic stage is past and the organ has become firm and indurated like cartilage, massage and all other remedies will be useless. The author makes a noteworthy distinction in pointing out the fact that the pains and discomfort accompanying enlargement of the uterus really are seated in the walls of the abdomen, though usually referred to the uterus; and these are first subjected to massage,

---

[1] Such innocence is a fair sample of the interest, or the want of interest, taken in massage. I was amused at seeing a part of one of my own papers used as an original article in a western medical journal without the quotation marks, and again quoted from this by a journal in New York where it was first published.

gently and superficially to begin with, then more deeply and vigorously until sensitiveness lessens sufficiently to allow the uterus to be kneaded. If this cannot be done effectually through the abdominal walls, "the first and second fingers should be passed into the space behind the vaginal portion, which is pulled gently forward, and then permitted to return to its former position. This is repeated a half dozen times or more, when the fingers are pushed higher up, so as to reach the supra-vaginal portion of the cervix and lower part of the body. The upper part of the uterus being now steadied by the hands on the outside, it is pressed between the fingers of both hands, repeatedly, for a few seconds at a time, and then relaxed. Every portion of the organ which can be reached should be subjected to these momentary squeezings. Then the manipulations should be reversed. The intravaginal fingers should be drawn in front of the cervix, and the latter pushed backward several times as far as possible short of causing pain. Then, their ends being passed into the space between the bladder and the cervix, and their pulps turned against the latter, the fingers of the outside hand should be so adapted that the uterine body may again be brought between the compressing forces, when the squeezing and imparted movements are to be repeated as before. Alternating with the process described, the uterus should be frequently elevated in the pelvis and held for a few seconds." After this the details of several cases are given in which benefit came from this mode of procedure.

When the hyperplasia is dependent upon some local condition outside of the uterus, such as inflammatory exudations in the pelvic cellular tissue, and when spots of tenderness and indurated fibrous bands are found fixing the uterus in some abnormal position, the enlargement will remain so long as the surrounding induration continues, owing to the disturbance of the circulation which passes through the cellular tissue in going to and coming from the uterus. The removal of these conditions external to the uterus must be secured by appropriate means, before any diminution can be obtained in its size. Amongst appropriate means massage is not suggested by Dr. Jackson. It is just here that this treatment has been taken up and used effectually by another independent worker, Dr. Otto Bunge, of Berlin.

In the *Berliner Klinische Wochenschrift* of June 19th, 1882, Dr. Bunge has published an article reporting favorable results from massage of the abdomen, particularly of the uterus and its surroundings. In case of atony of the intestines with constipation, he advised patients to *masser* their abdomens themselves, and with advantage. He does not fear the apprehensions of Von Mosengeil, that perforation might be caused in this manner by the possible existence of ulcers, so long as the massage is performed with proper gentleness and never in recent affections. He has used massage most frequently for the removal of the sequelæ of peri-uterine cellulitis and pelvic peritonitis of the most various forms, which had made defiance to the customary methods of treatment. When engorgement of the uterus was also present, as in subinvolution and movable retroflexion, this method proved excellent, as may be seen by some of his tabulated cases. His mode of using massage was very much similar to that of Norström and Jackson. But as his aim was often the loosening of adhesions and the dispersion of indurations, the manipulations in such cases were directed towards the seat of these, working more around the uterus, internally and externally, and pushing, pulling, or raising it in such ways as would detach the adhesions. The good effects of this treatment showed themselves by the dispersion of the pathological products, thus increasing their surface for resorption, by furthering the circulation and by stirring up the contractions of the uterus. One patient declared that while being *masséed* she felt real after-pains, although four years had elapsed since her last confinement. Cases were treated in which the uterus was so closely fixed to one or the other part of the pelvic walls by adhesions that at first it was not possible to penetrate between them. With these only gentle steady pulling or pressure could be used, but by patience and perseverance " they became the most thankful of all cases." Dr. Bunge soon learned that precaution, tact, and skill were necessary, and he candidly confesses that he treated his first cases, which are also reported in his table, too heroically. Injections of warm water, which he also uses at first with manipulation, act very much in the same manner as massage, but not so effectually in stretching, loosening and further resorption. When there are inflammatory products in the cul-de-sac of Douglas, or in the perivaginal tissue, the author says that less benefit proceeds from the usual local medication than from the mechanical pressure made in applying it by means of the speculum which produces resorption; but it is better to have these products under the sensation of touch. Contra-indications of massage in these cases would be those pathological conditions in which purulent or sanious products are present or even suspected. In concluding, Dr. Bunge strongly recommends massage for the removal of old perimetritic and parametritic sequelæ and congestion of the uterus; and from the results gained in Nos. 14 and 17 of his table, he promises a great future for massage in the treatment of flexions of the uterus without instrumental aid. In the time that had elapsed from the presentation of his paper to its publication, cases which were not quite cured had got well, and others had come under treatment.

DR. OTTO BUNGE'S CASES.

NO., NAME, AND AGE.	CONDITION.	DETAILS OF TREATMENT.
1; Gr.; 31 years.	Parametr. chron. dextr. Endometr. Parametritic thickenings fix the uterus towards the right and backwards.	Only *masséed* a short time. During the treatment fresh parametritic irritation, which came after brisk exercise and sitting on the cold ground. Disappeared from treatment.
2; H.; 21 years.	Perimetr. et parametr. sin. chron. Fornix vaginæ very sensitive on the left side, where are firm adhesions. Both sacro-uterine ligaments thickened and shortened.	Treatment instituted at another place; continued by me, viz.: preparations of iodine, sitz-baths, cataplasms, and injections without benefit. After trying massage a few times the patient left.
3; J.; 31 years.	Retroflex. uteri fix. Parametritis chron. sinistra; later dextra. Uterus retroverted and held fast by perimetritic adhesions, especially towards the left side and behind.	Preparations of iodine and cataplasms, with which the patient was treated for 3 months, were not well tolerated. Massage once or twice weekly for several months caused decided improvement. The remains of the exudation almost entirely disappeared, general condition became excellent, and weight increased. This patient

NO., NAME, AND AGE.	CONDITION.	DETAILS OF TREATMENT.
		was stepped on by a boy, which set up fresh parametritis on the right side in March, 1881. In acute stage antiphlogistics; then hot water and Priessnitz. From Oct., 1881, to Dec. 1st, the uterus was fixed by adhesions. After eight massages the uterus is in the upright position, and the patient is satisfied.
4; Z.; 33 years.	Chronic parametritis, with adhesions. Uterus fixed and retroverted. Treated elsewhere for several years without benefit.	From the middle of September, 1880, to the 1st of January, 1881, astringent baths and injections, iodoform suppositories, rectal injections, etc. After ten sittings of massage, begun in January, the uterus could be brought in a forward position without material pain. After the twelfth massage in April, the patient was not seen till September. On account of profuse and continuous menstruation, the fungous mucous membrane was scraped, and after this the progress was uninterrupted.
5; Sch.; 37 years.	Parametr. chronic dextra. Uterus retroverted and fixed. Adhesions from the posterior wall of the uterus to the pelvis.	Treated at another place by pencillings, sitz-baths, Priessnitz, etc. Also by myself; at first treated for awhile, with immaterial improvement, by means of iodoform-vaseline tampon, painting the cervix with iodine, sitz-baths, and cataplasms. Since the beginning of October massage two or three times daily. After a few weeks the adhesion disappeared.
6; Gr.; 25 years.	Endometritis and chronic perimetritis. She has three times aborted—at 6 weeks, 3 mos., and at 4 weeks respectively. Uterus not enlarged, but moved only with difficulty, and distorted backwards and to the left. Posterior part of the fornix vaginæ very sensitive. At the posterior portion of the tissues on the left of the uterus there are strands of exudation. Slight eversion of the cervical mucous membrane.	From the end of August to the middle of September, 1880, painting with iodine and rinsing out the vagina with solution of iodine, salt baths, and a tonic of calisaya bark. At the end of July, 1881, condition but tolerable. Appearances as upon first examination, except that the cervical portion is much more swollen, and there is greater eversion of the mucous membrane. Menses at last flow of bad odor. From the 21st of July, 1881, to the 2d of August patient was masséed daily, and in the evenings she had injections of hot water. The menses then came easily, and without bad odor. General condition much improved. Massage daily from the 24th to 31st August. Cured.
7; H.; 29 years.	Chronic parametritis following abortion several years ago.	Formerly treated elsewhere. A few small adhesions and thickening of the sacro-uterine ligaments disappear after two sittings of massage,

NO., NAME, AND AGE.	CONDITION.	DETAILS OF TREATMENT.
8; K.; 32 years.	Perimetritis. Anteflexion. Chronic endometritis. Ectropion of the os uteri; erosion. Besides several adhesions there are separate nodules, from the size of a coffee bean to that of a hazelnut, in the cul-de-sac of Douglas. Cervix swollen. Inclination to fungus.	and the motion of the uterus is improved. Elsewhere many times treated with iodine injections, sitz-baths, etc., without improvement. Only massage and hot water used by me. After seventeen massages, uterus tolerably movable; erosion disappeared; still slight thickening in the space of Douglas and upon the anterior wall of the uterus can be felt. Positive cure after twenty-seven sittings.
9; Pfl.; 24 years.	Chronic perimetritis and parametritis. Endometritis. Abundant granulations on the anterior lip of the cervix. Uterus held firm as a wall by the mass of adhesions, especially at the sides and behind.	Patient masséed her bowels herself with good results. The massage of the uterus also made good progress. Granulations removed by lapis. The uterus became more moveable from week to week, and by the 1st March it was entirely free. Massage suspended on account of trouble with stomach (ulcer or carcinoma?)
10; B.; 39 years.	Chronic pelvic peritonitis on both sides of the uterus, and also around the vagina and rectum. Tissues infiltrated to a great extent.	October 1st, 1881, the cavity of Douglas opened, and pus evacuated. To the middle of December, rest, hot water injections, and Priessnitz. Uterus was firm as a wall by masses of adhesions all around. Massage twice a week. Patient improves and gains in weight. By the middle of February, there was slight mobility of the uterus. By the 1st of March left side almost free, but the right still quite adherent, which, with the parametritis behind, holds the uterus firm.
11; V.; 24 years.	Chronic parametritis on left side. Anæmia, with too much obesity. Aborted two years ago. She now imagines herself pregnant. Abdomen very flat. Suffers from chronic constipation. To the left, parametritis has thickened and shortened the sacro-uterine ligaments. Uterus moved with difficulty.	After ten massages the thickening had disappeared, and the uterus was movable without pain. The bowels became regular in the most satisfactory manner by her own kneading of them. Baths and exercise in the open air should remove the anæmia.
12; St.; 27 years.	Anæmia. Subinvolution of the uterus after abortion. Uterus relaxed, retroflexed, enlarged, and movable.	After hot-water injections, the cervix underwent involution very well, only the body was relaxed and enlarged, and showed an inclination to bend backwards. Two sittings of massage brought the uterus to a good state of

NO., NAME, AND AGE.	CONDITION.	DETAILS OF TREATMENT.
13; B.; 29 years.	Ut. retroverted. Chronic parametritis on the right side. Uterus fixed to the right and backwards by abundant and tolerably firm adhesions.	contraction and to an anteflected position. Massage at first painful, but became less so. After five sittings, subjective improvement and appetite increased. Upon traction with bullet forceps in the cervix, the uterus followed, with slight crepitation. Manual reposition. After the ninth massage only slight thickening of the right sacro-uterine ligament. Cured by thirteen'sittings of massage.
14; Schl.; 29 y'rs.	Retroflexion of the uterus, with slight parametritis. Uterus both thick and long. Upon its posterior surface and in the cavity of Douglas many thin, firm deposits can be felt.	Massage only painful at the beginning of the first sitting. Upon the disappearance of the deposits in the cul-de-sac of Douglas and in the posterior wall of the uterus, the uterus resumed the upright position. After ten sittings, the uterus was small, and remained a whole day anteflected, and then fell back again. Cure was interrupted for several weeks by a journey for health, during which a pessary was worn.
15; H.; 27 years.	Endometritis. Metritis and chronic parametritis on left side. Rupture of the cervix. Erosions. Eversion of the mucous membrane. Delivered by forceps 5 months previously by a colleague.	Injections of hot water and tannic acid without result. Even after three massages the secretions had considerably diminished, and the erosion of the cervix, now well contracted, had disappeared. At the fifth massage, the uterus was free from adhesions, anteflected, and could be treated without pain.
16; Schm.	Chronic parametritis. Uterus fixed anteriorly and posteriorly. Complains of involuntary dribbling of urine (pressure of the anterior fixed portion upon the urethra and under part of the bladder). Uterus stands about in the right diameter of the pelvis, but slightly bent backwards.	Even after two massages the patient could retain her urine for hours at a time, and after four massages the ends of the fingers could be placed perfectly well between the symphysis and the uterus, both by the vagina and through the abdominal walls. The adhesions posteriorly were also disappearing. After nineteen sittings, free from annoyance and discomfort. Still both sacro-uterine ligaments remained thickened and shortened, and behind the right adhesions can be felt. Patient still under treatment.
17; K.; 32 years.	Uterus hypertrophied and retroflexed, the size of a small fist, and moveable. Wore Hodge's pessary from August 23d, 1881, to February 1st, 1882.	Pessary removed on account of bad odor. After twelve days, uterus about as in first examination. Patient now treated by massage, and felt considerable after-pains during manipulation. After five days uterus kept in a desirable anteflected position for four days; after the sixth sitting in normal position for six days; and after another massage in normal position for seven days.

Dr. Bunge warns physicians against considering massage a panacea for all sorts of uterine affections.

Winiwarter, in the *Wien. Med. Blätter*, 29–31, 1878, reports the case of a woman, seventy-nine years of age, suffering from a multilocular ovarian cyst with consequent œdema of the lower half of the body. After repeated aspirations of the cyst, the fluid always accumulated in greater quantity than before. The patient declined an operation for its removal. Winiwarter concluded to try massage upon the legs, and this soon removed the swelling, relieved the pain, and increased the urine. Later he extended the massage upon the abdomen, and over the ovarian cyst, and this decreased in size and remained smaller. Œdema did not return, and the condition of the patient was improved. It is noticeable that when W. was at one time hindered from repeating the massage for a while, the former condition of the patient returned. The difficulty, however, soon disappeared, the swelling becoming smaller, when Winiwater resumed the massage himself. He emphatically guards himself against being open to the accusation of believing massage to be able take the place of ovariotomy or aspiration (as it did in this case); but in cases where ovariotomy or aspiration is impracticable or refused, massage may be of use. In his opinion, the intermittent pressure of massage acts more powerfully than permanent or continued pressure.

Dr. H. P. Orum has found massage efficacious in removing the after-effects of periuterine cellulitis when other means have failed. The infiltration of the connective tissue after parametritis is, in the majority of cases, comparatively speedily absorbed; but in a few cases it remains for a long time. At Professor Howitz's clinic, massage is used successfully in these cases for five or ten minutes every other day, after all other therapeutic measures have been exhausted.

In the *Journal de Médecine* of January 3rd, 1886, we learn that Dr. Prochownik has used massage in 103 cases of chronic uterine trouble. Sixteen of these could not go on with the treatment on account of the pain due to the inexperience of the operator. Five of the remainder were cases of small intra-ligamentary tumours, and two of these disappeared under massage and did not return. Of 13 cases of prolapsus of the uterus only one was cured and two improved. Of 10 cases of chronic metritis four were cured and three improved. Of 18 cases of exudation, including five of hæmatocele, eight recovered, two improved and three were slightly benefited. (What became of the others?) The most suitable cases for massage were found to be those of old cicatricial, contracted remnants of exudation, and amongst 40 of these 24 were cured and 10 greatly improved. In 10 cases of latent gonorrhœa, painful joint affections followed, associated with slight fever; so that Prochownik was led to regard this as a contra-indication for massage. Other contra-indications are pregnancy and consumption. In addition to manipulation with the hands and fingers in the region of the uterus he also employed what he calls passive massage in some of the cases. For this purpose a series of vulcanite cylinders were employed in order to gradually dilate the parts contracted by cicatrix or spasm.

According to Prof. Bartholow, electricity is only of value in uterine disorders when there is *no* hyperplasia of the connective tissue. According to various observers massage is of great value when there is hyperplasia of the connective tissue.

More recently others have had favorable experience with massage in gynæcology. By this means internally and externally, and pushing the uterus upwards and forwards and allowing it to fall back again, varied with resistance to adduction and abduction, Prof. F. von Preusschen [1] benefited a case of prolapsus uteri of 31 years' duration so much that after the first day of treatment the uterus remained in the pelvis, and at the end of 3½ months there had been no prolapse. The duration of treatment is not stated. The raising of the uterus is of value in correcting the retroversion and for separating adhesions between the uterus and bladder. The resistive movements of the thighs cause the whole muscular floor of the pelvis to contract, and this can be increased by the patient raising her hips at the same time. In this way the lax muscles regain their tone, they re-

---

[1] Centralblatt für Gynœkologie.

store the proper support to the cervix and narrow the opening for the vagina in the pelvic floor.

By like procedures Dr. Paul Profanter[1] cured a case of prolapsus of 27 years' duration in 10 days, and another case of complete prolapsus of 10 years' duration in a month. In the last case the uterus remained 5 cm. above the perineum after the first treatment. Besides these he has used massage in 14 other cases of parametritis with abdominal fixation of the uterus, chronic ovaritis and periovaritis, fixed retroversion and one case of retro-uterine hœmatoma. Abnormal fixations and positions were removed within a few weeks, and the accompanying pains and discomforts also disappeared. The conditions which were found by careful palpation before and after treatment are shown by wood-cuts. The results were surprisingly favorable.

An eminent writer has styled massage an agent of singular utility. Its effects in such cases as the above show it to be an agent of plural utility — in the dispersion of morbid products, in the loosening of adhesions, in correcting local and general disturbances of circulation, and morbid secretions dependent upon these, in restoring contractility and tonicity to organs surrounded by involuntary muscular fibres, as the uterus, the bladder, and the intestines, and thus enabling them to resume their normal functions. Massage of the uterus and the tissues adjacent to it, first brought prominently to the notice of physicians by a layman, has found its place in rational therapeutics through the instrumentality of Norström, Asp, Jackson, and others. Norström seems to have been the next after Dr. Sinclair to use massage scientifically in these cases, the nature of which could be understood, and the mode of action of massage explained. The beneficial effects of general massage upon local affections in women are shown by the experience of Asp, and others before him, mentioned in the preceding chapter. Independently of any one else, Dr. Jackson has neatly combined the advantages of internal and external massage so as to gain the best effects with the least effort, and just where he has left the increasing range of massage of the pelvic organs, Dr. Bunge has

---

[1] Die Massage in der Gynœkologie, von Dr. Paul Profanter, Vienna.

taken it up and proved at one and the same time its good effects in the class of cases in which Jackson found it beneficial, and still further for the removal of indurations and adhesions around the uterus. Besides the experience of Prochownik, Von Preusschen, and Profanter here given, the medical journals now make frequent mention of the results of massage by others well qualified to judge ; so that this method, which a few years ago was sneered at, has gained for itself a fixed place in the treatment of intractable uterine affections as well as in many other conditions. Some physicians who are either too busy or too modest to record their results, or who fear that they would be doubted, have told me of the excellent service that massage affords them in their own hands in uterine maladies, which they can obtain in no other way. Still other methods are used for the same purpose, such as packing the vagina, gradual dilatation, etc., all of which are more or less imitative of massage. But skilful gynæcologists are not very likely to apply massage until everything else has failed, for it is tedious and difficult for the operator, painful and disagreeable to the patient.

CHAPTER VII.

THE EFFECTS OF MASSAGE UPON INTERNAL ORGANS, CONTINUED.

UPON SIMPLE CHRONIC HYPERÆMIA OF THE LIVER; UPON ATONY OF THE
STOMACH AND INTESTINES; UPON INTESTINAL OBSTRUCTION, AND UPON
PERINEPHRITIC INDURATION WITH SEVERE NEURALGIA, ETC.

" Last, with a dose of cleansing calomel
Unload the portal system — that sounds well."

CONTINUING a consideration of the effects of massage upon internal organs, we can easily see that the ascent and descent of the
diaphragm in respiration, and the influence of exercise in general,
keep up a sort of perpetual massage of the liver, which materially
aids its functions and preserves the equilibrium of its circulation.
Loosely suspended beneath the diaphragm by folds of the peritoneum, the situation and attachments of the liver are peculiarly
favorable to allow it to gravitate in any direction, or to be pushed
upwards by full intestines or downwards by dilated lungs, to be
drawn upwards by the arched diaphragm of phthisis or to gravitate downwards when the intestines are empty. While in health,
natural movements and ordinary exercises suffice to keep the
liver in good condition, yet when this organ becomes hyperæmic,
these are of necessity often lessened and insufficient. At every
meal there is an increased flow of blood to the liver, and there
may be too great a determination of blood to this organ from over-
feeding, the portal vein itself being too much filled, and all the
digestive organs overtaxed, to overcome which the kings of the
Sandwich Islands have themselves *lomi-lomied* after every meal.
When this state of affairs has been too long continued without
anything to counteract it, there may result hepatic engorgement,
the treatment of which by the douche and massage has been suc-

cessfully carried out by Dr. Durand-Fardel, Director of Vichy.[1] According to this writer, these are valuable adjuncts to other treatment. Simple hepatic engorgement or chronic hyperæmia of the liver is indicated by general or partial enlargement of the organ, usually of the left lobe. There are no tumors or inequalities of surface, nor excessive hardness. Pain is only observed at intervals, whether spontaneous or aroused by pressure, and is at times wanting. It is neither of the character nor fixity of cancerous pain. Icterus may be absent or faint. There is no cachectic look, as in organic affections. Neither the glycogenic nor hæmatosic functions are at fault. Anasarca and ascites are absent. Simple engorgement of the liver is susceptible of resolution after long duration. Its persistence may lead to the belief that hypertrophy or induration has replaced it.

Of 133 cases noted by Dr. Fardel, 43 were said to have followed upon acute attacks resembling hepatic colic ; 72 developed gradually, and 18 in a latent manner, their advent being masked by dyspeptic symptoms. Whatever the method of onset, this sort of engorgement, when fully developed, presents nearly the same symptoms in every case. It may be connected with cardiac disease. The time-honored treatment by Vichy has comprised simply baths with the internal use of the waters. Fardel has added to these douches and massage over the hepatic region. The douche is in the form of a fine spray, and is used five or ten minutes at a time, every other day to begin with, at a temperature of 93° F., and just before the bath. It is followed by a general feeling of buoyancy and of relief in the hepatic region. While the douche itself is a sort of massage, yet there are cases where massage may be also employed. Then the abdomen is first kneaded altogether, after which the hand is gently passed over the hepatic region, the skin of which is at first squeezed and afterwards the deeper parts are worked, and this is alternated with tappings by means of the palmar surface of the fingers, and finally the liver itself is kneaded, its lower edge being raised and seized by the hand. These manipulations are done with great gentleness and gradually accomplished, not being finished at the first sitting. The operation takes from

---

[1] Bulletin Gén. de Thérapeutique, Vol. c., 1881, p. 241.

five to twenty minutes, and is repeated every other day. Buoyant feelings result, and the massage is agreeable while being done. Intercostal neuralgia is not a contra-indication for massage, as it is for the douche. By massage the liver is partly emptied of its fluids, the circulation is favorably acted upon, and absorption hastened. In most cases this treatment is satisfactory, the general condition of the patient improves, and the dyspeptic symptoms disappear. The liver does not always diminish in size during the month or so that the patient is under treatment, but the diminution and general improvement go on after the patient leaves. The entire treatment of these cases lasts for several years, and what is gained at each visit is retained, and in the great majority of cases the patients ultimately recover.

It will be observed that the careful manner of applying massage to the liver by Fardel is as much like that of Jackson in applying the same treatment to the uterus as it possibly can be. The position of the patient, the state of the stomach and bowels, and the time at which massage may be used to the best advantage for hyperæmia of the liver are not spoken of by Fardel. For such a purpose, I prefer that the time shall be long after a meal, when the stomach and bowels are comparatively empty, and that the patient shall be sitting, with the body inclined forward, resting upon the elbows, so as to relax the abdominal walls and allow the liver to gravitate downward and forward, and this will be greatly aided by the patient making gentle, deep, and prolonged inspirations.

In poorly-nourished people who gain flesh under massage and feeding, no small part of their improvement may often be due to the effect of massage over the region of the liver by which its functions are stimulated; for the bile aids the pancreatic juice in emulsionizing fats and preparing them for digestion, besides its presence in the intestines excites their peristaltic action, as does the immediate mechanical effect of massage itself. Many have, no doubt, heard of the invalid lady who was suffering from constipation and, as a result of this, indigestion. Her friend, calling one afternoon, inquired : " Do you ever knead your bowels ? " The invalid meekly replied : " Indeed, ma'am, I cannot very well do without them." The resulting burst of laughter shook up the abdominal viscera to such an extent that the constipation was relieved, the indigestion was cured, and the patient got well. There is truth in the old adage: "Laugh and grow fat." "Shake up the chylo-poetic viscera," is the doctor's language for the same, and hence he often prescribes horse-back riding. Obstruction of the excretory gall-ducts from catarrhal swelling or collection of mucus might sometimes be relieved by massage squeezing out the contents of a distended gall-bladder, and thus pushing the obstruction before it. In the disappearance of icterus and sudden improvement following examinations of the liver, this is probably what has occurred.

Atony of the muscular coat of the stomach or intestines with deficient peristaltic action and consequent disturbances of digestion, accompanied with distention from flatulence or solid contents, is usually benefited to a marked degree by appropriate massage, after ordinary exercise and other measures fail. That benefit is more likely to follow here from repeated treatments than as an immediate effect would show that the nerve-centres that preside over these functions have undergone a nutritive change which has taken time to produce, and hence that the improvement would most likely be lasting. Moreover, when the alimentary canal is distended by gas or overburdened by solid contents, the nutrition of its walls must suffer from languid circulation, as any muscular organ would that was continually stretched and inactive. Massage improves the circulation and pushes along the contents of accessible portions of the stomach and intestines at the same time, besides directly stimulating the muscular fibres to contraction, and reacting on the nerve-centres, thus improving function and organization in various ways.

For habitual constipation Dr. Sahli of Bern advises his patients to roll a five-pound cannon-ball upon the abdomen for five or ten minutes every morning before rising. In this way he has cured nearly all his cases of torpid bowels without medication. When universal peace comes, the orator can therefore speak not only of turning swords into ploughshares, but also of cannon-balls into aperients; and peace will then have its victories no less renowned than war.

In dilatation of the stomach and chronic dyspepsia massage has proved of great benefit, the distressing symptoms disappearing, and the patients gaining flesh and strength. It should be applied as much as possible over the stomach, working from the left side upwards and inwards, under the false ribs, so as to empty the stomach of its contents, whilst stimulating the contractility of its muscular walls. The effect of gravity causes the food to lodge at the most dependent portion in the greater curvature, and it needs to be pushed along towards the pylorus, whilst the peristaltic action is being increased. The whole abdomen should be *masséed*, at the same sitting, as the bowels are usually sluggish in these cases. Rubens Hirschberg, of Odessa, reports numerous cases of this kind which he relieved or cured by massage, and lays special stress on the fact that massage has an ultimate chemical influence in causing the disappearance of sour, burning eructations, fetid breath, and bad taste in the mouth, as well as sensations of weight and fullness, which would indicate an improved state of the gastric juice and better contractions of the stomach. Hirschberg invariably found in these cases that massage of the abdomen for 30 minutes increased the daily quantity of urine, in some cases to three times the usual amount, without any other inconvenience than the necessity of frequent micturition. When massage was discontinued the urine fell to its former quantity. The increase of urine probably depended upon more active absorption of fluids from the digestive tract, increased blood-pressure, and stimulation of the splanchnic and pneumo-gastric nerves below the diaphragm caused by the massage. Contra-indications to the use of massage upon the stomach would be symptoms of cancer or ulcer, acute or febrile states and suspicions of a tendency to hemorrhage. If cicatricial contraction of the pyloric orifice existed, massage might increase the dilatation, for the walls of the stomach would in all probability yield more readily than the cicatrix.

But except in thin people massage of partly accessible internal organs, such as the liver, the stomach and uterus, is neither so easily nor so effectually accomplished as many try to make out.

In order to obtain a mixture of the pancreatic juice, the bile and the succus entericus, Dr. J. Boas, of Berlin, rubs the abdomen from the region of the right hypo-chondrium towards the median line, after the stomach digestion has ceased, and that organ is empty. This procedure gradually relaxes the pyloric sphincter, and after a time the intestinal juices enter the stomach in considerable amount. These are then withdrawn by the stomach-pump. The amount of secretion thus obtained in twenty cases was on an average from 40 to 50 cc. for each.

Seventeen years ago the writer *masséed* a patient who suffered from emphysema of the lungs, together with obstinate constipation. · Besides general massage, there was given special massage of the abdomen and liver, with percussion over the latter and pressure upon the chest-walls during expiration. Under this treatment alone, the stools, from being of a pale color, became natural and occurred twice daily, respiration was easier, sleep and appetite improved, and the general condition was much better.

Straining during forced inspiration, as in lifting heavy weights, or in difficult defæcation may be an exciting cause and, by its continuance an aggravation of pulmonary emphysema, in those predisposed thereto. The pulmonary circulation, obstructed by the dilated air-cells, the dilated hypertrophy of the right ventricle, the consequent hindrance to the outlet of the circulation from the liver, and backing-up of this in the portal vein, will only in part be relieved by massage or any other means that will facilitate and increase the area of the systemic circulation. In the case just mentioned, ordinary exercise aggravated the symptoms.

As a necessary preliminary to the solution of biliary calculi, fracture of the crystals must occur. This is often effected on the part of nature by the movements of respiration and of the abdominal muscles, which cause more or less attrition, and breaking of the edges and corners of the crystals, and thus permit the solvent action of the bile. This process may be aided by manipulation of the gall-bladder through the abdominal walls. Faradization has been recommended for this purpose, but it must be evident that massage carefully applied would do this much better. For attempting the disintegration of an impacted biliary calculus, Prof. Bartholow recommends that firm friction be made with the fingers along the inferior margin of the ribs, and towards the epigastrium and umbilicus, whilst the opposite side posteriorly is supported by the hand spread out and firmly applied.

Several cases of pleuritic effusion, where the absorption of the fluid was accelerated by percussion upon the chest-walls, have been reported by Dr. Emil Schlegel in the *Allgemeine Med. Central Zeitung*, No. 20, 1885. For this purpose the ulnar border of the hand was used, striking at the rate of two blows a second, or 600 in five minutes. Two *séances* were given daily. Schlegel believes that percussion might be quite as useful for promoting absorption in other parts of the body, as the intra-cranial cavity, spinal canal, etc., which are not directly accessible to manipulation. But as the walls of these cavities are unyielding, the blows would have to be so strong that the patient would probably object.

At a meeting of the Royal Medical and Chirurgical Society, a report of which appeared in the London *Lancet* of December 18th, 1875, Dr. Brinton pointed out that of six hundred cases of intestinal obstruction, forty-three per cent were due to intussusception, and of these thirty to forty per cent terminated favorably. Dr. Brinton stated that the operation of gastrotomy for the relief of these cases is only of value in the earlier stages of the affection, when the condition is one of obstruction and not of enteritis. The question was very appropriately asked whether such a dangerous operation as gastrotomy was justified in a disease of which the proportion of cases of recovery by other means is so high. A full account of a case of "intestinal obstruction of five days' duration cured by kneading after injection per rectum" is given in the London *Lancet* for July 27th, 1872. The patient was an adult and a free liver. To the right of the umbilicus, and above it, there was distinct hardness which gave the impression that a transverse coil of the bowel could be felt above bending upon a vertical one below. The vomited matters were brown, with flocculi of a darker color, but not stercoraceous. Anodynes, hot fomentations, and injections had been thoroughly used, but no relief was obtained until the abdomen was kneaded by Surgeon Brookhouse, of Deptford, who attended the case. The reporter, Dr. Fagge, remarks: "It can hardly be doubted that the life of the

patient was saved by the kneading of the belly, and so satisfactory
an issue may well encourage other surgeons to adopt a similar
procedure. Yet it cannot be denied that too forcible manipulation
of the abdomen might, in many instances, involve great risk of
tearing through parts softened by inflammation or sloughing, and
thus counteract the curative processes of nature. There was
nothing in the symptoms of the case that indicated the special application of kneading." According to Dr. Fagge, massage saved
this patient's life when there were no indications for it. What,
then, will it do when there are indications for it? Certainly no
one of any sense would think of using massage if it could be
made clear, or even if there were any doubts, that enteritis had set
in or that there was danger of sloughing. The manner in which
massage was used in this case is not stated, but when the condition
is one of obstruction alone, what mechanical agency can be more
likely to disperse a mass of fæces and push it through a constricted opening, or pull out a portion of invaginated intestine than
massage suitably applied? It seems to me that the indications
for kneading at this stage are special, and that the best manner of
doing massage here would be by gentle stroking and kneading at
a short distance beyond the distal or rectal end of the mass, and
in the course of the intestine towards the anus, continuing to
work in the same direction, and gradually proceeding backwards
upon the tumor. In this way, the intussusception would be most
likely to be pulled out, the mass of fæces broken up and pushed
along in their natural course. Until this milder procedure has
been thoroughly tried, no patient should be vivisected, for it is
only in the same stage of the affection that either massage or
surgical interference might be effectual. Massage has been effectual in every instance that we have heard of, but it may have
failed in a great many more that we have not heard of. How
often has it been tried?

Dr. C. P. Putnam[1] has reported a case of intussusception
of the large intestine in a child five months old successfully
treated by injections and massage. A cylindrical tumor, about
three inches wide and one and one-quarter inches in diameter, was

---

[1] Boston Med. and Surgical Journal, April 21st, 1881.

felt between the umbilicus and left costal cartilages. Gentle massage was applied to the tumor with the intention of diminishing the hyperæmia and œdema, and possibly of reducing it. Under this treatment the tumor became softer and shorter. Water was then injected into the bowel. This was at 10 P. M. In the morning the patient passed three bloody stools, and the tumor was again found as on the previous evening. A repetition of the same procedures was successful, and the patient recovered.

Buch[1] has used massage in four cases of intussusception successfully, the patients recovering. One was a case of strangulation at the ileo-cæcal valve, whilst the other three were regarded as invaginations. In these conditions it is considered impossible to fix the place of strangulation unless the tumor formed by the arrested matter can be felt. Then massage is used from above downwards in the track of the intestine, at first attempting to push the fecal mass past the constriction, and then stretching so as to remove the invagination. When the seat of the constriction cannot be located, it is not a contra-indication for the employment of massage. It is always advantageous to displace the fecal mass and to remove it to another part of the intestine. By this means it is broken up into small parts, so that purgatives succeed better in expelling them by peristalsis and hyperexcretion. It is impossible for an intestine greatly distended to contract.

One of Buch's cases was that of an inn-keeper who had always been in good health. No movement of the bowels had taken place for four days. The patient was suffering violent abdominal pain, alternating with vomiting, there was meteorism and dyspnœa, and the countenance was expressive of great anxiety. Above the umbilicus on the left, there was a swelling in the form of a large pudding, having its convexity directed upwards. Massage was used so as to propel the anal extremity of this onwards, and then was gradually extended further upon it until no trace of it was left. Several hours afterwards the patient had a copious dejection (*une riche garderobe*) and the next day was up and well. Another of Buch's cases was that of a lady who had not had an evacuation of the bowels for thirteen days, and who vomited everything she

---

[1] Berliner Klin. Woch., October 11th, 1880. Norström, Du Massage, 1884.

took.   There was no hernia nor swelling.   An œsophageal sound
was introduced by the rectum so that it could be felt in the
epigastric region, then an abundant injection of ice-water was given,
which came away in two hours, but no fæces with it.   Another
examination revealed a moveable elongated tumor commencing at
the right inguinal region and extending towards the left.   It seemed
probable that this was the inferior part of the small intestine and
that there existed a constriction at the ileo-cæcal valve.   The walls
of the abdomen being thin, there was no difficulty in pushing along
the distal extremity of the mass, and this was *masséed* until no
more of it could be felt.   With the aid of aloes she had her first
dejection thirteen hours after massage, and recovered.

In *L' Union Médicale* of 18th March, 1882, Bitterlin reports two
cases of intestinal obstruction with vomiting of fecal matter which
were treated with massage and got well.   The author concludes by
saying that he " considers it his duty to publish these cases in
order to show that in obstruction of the intestines massage of the
abdominal region can bring about quite unlooked-for results when
other means have failed.   Before having recourse to such extreme
measures as puncture of the intestines, enterotomy or gastrotomy,
operations which are always of a certain gravity, it is important to
try massage, which can effect cure in the most desperate cases."

Scerbsky, in the *Petersburger Med. Wochenschrift*, 1878, reports
the case of a boy, six years old, who had all the symptoms of an
invagination — tumor in the form of a pudding in the left iliac
region, violent pains in the abdomen, vomiting, meteorism, tenes-
mus, and collapse.   After having tried injections and purgatives
in vain, the intestine was punctured to let out the gas, then the
swelling was *masséed* with great care on account of the pain.   At
the end of ten minutes, something was felt gliding under the fingers,
and there was heard at the same time the gurgling of flatus and
the pain ceased.   Abundant vomiting followed, then the child
went to sleep, and when he awoke all the unfavorable symptoms
had disappeared.   During the night there were copious stools,
and cure resulted.   Krönlein reports like favorable results from
massage in intestinal obstruction.   (*Reibmayr*).

Dr. Kriviakin warmly advises deep massage of the abdomen as

144

a powerful curative means in cases of intestinal obstruction. He has reported four cases in which the symptoms were obstinate constipation, agonizing paroxysmal abdominal pain, fetid vomiting, obstinate hiccough, offensive eructations and distention of the abdomen. *Séances* of fifteen minutes every hour and a half were employed. As a rule one sitting was sufficient to produce stools. Recovery resulted in three of the cases; and the fourth was that of a decrepit, weak man, regarded as hopeless before massage was tried. But notwithstanding this, in about half an hour after the first massage there was a free discharge of fecal lumps suspended in fluid. The patient was in a semi-comatose state, with cold, viscid perspiration, fetid vomiting and eructations, filiform pulse with constipation of twelve days' standing before massage was begun. After massage collapse became worse, and five hours later the man died. No autopsy. According to Kriviakin, deep massage is indicated in intestinal obstruction of every kind and description. [1]

A case of induration of the cellular tissue around the kidney with rebellious neuralgia of the leg, the sequel of perinephritis, was cured by massage at the hands of Dr. Winiwarter. The patient was a vigorous corpulent man fifty-eight years of age, whom Prof. Lobel sent to Winiwarter to be treated with massage. He had suffered from violent neuralgic pain in the left leg for five years, and this had resisted all sorts of treatment, and for the last two years his life had been passed between his bed and table, and he was obliged to keep absolutely quiet as much as possible, for he could only repose lying on his back. Sitting, standing, or walking precipitated paroxysms of pain which even occurred without change of position to the number of sixty a day. He could take a few steps supported by a cane and and by one of his domestics. The pain radiated over the external surface of the thigh to the knee, and it was also felt in the toes. Except wasting of the muscles of the limb and hip, nothing was observable; pressure upon the sciatic nerve at its exit did not excite pain. Over the region of the left kidney there was found a flat, uneven swelling which was difficult to palpate on account of its deep situation; it was firm and elastic, and over its middle pressure afforded an obscure sense of fluctuation. It was not sensitive to pressure except at a clearly defined point (not stated where), pressure upon which excited the severe neuralgic attacks in the thigh. The skin covering the swelling, though thick with adipose, was not adherent nor altered. Five years before W. saw the patient, the latter had suffered from a febrile attack with violent pains in the renal region, from which he recovered, but the pains persisted and assumed a neuralgic character. It was considered that the continuance of the trouble was owing to an exudation similar to what is found after periuterine inflammation, and that the pains were due to a compression of the lumbar plexus by this exudation, and that they were radiated from the lumbar to the sacral plexus. This view was concurred in by Prof. Billroth, who saw the patient. The treatment was daily massage of the limb, hip, and lumbar regions. At first this was painful, but in fourteen days the painful places had disappeared, and the patient began to take a few steps without support, and the paroxysms of severe pain occurred only once or twice during the day. After sixty-four days, the improvement was so great that he departed on his homeward journey. The tumor had then become very small, and fluctuation could not be felt; firm pressure upon its surface was still a little painful, but excited no attack of neuralgia. The patient took a walk of three hours every morning, and could get in and out of a carriage and stand upon the left leg. A melancholy state of mind had disappeared.

In chronic typhlitis and perityphlitis Dr. George Hämerfauth,[1] of Homburg, has found massage of great value. He points out that the cause of typhlitis is usually mechanical, the mass of fæces remaining too long in the cœcum on account of relaxation of its muscular coat, change of its position and chronic catarrh, against all of which massage would prove a valuable prophylactic as well as a curative measure. Of course this treatment has nothing to do with cases of abscesses or burrowing of pus. But for the removal of chronic thickening and adhesions affecting the cœcum and vermiform appendix, or their surroundings in the peritoneum and areolar tissue, the residue of typhlitis and perityphlitis, massage has been,

---

[1] Münchener Medicinische Wochenschrift, Mai 14, 21, 28, 1889.

in his hands, the most effectual agent. Indeed, his fifty-three cases treated in this way are sufficient to support him in the assertion that permanent improvement or recovery in severe cases of this sort can be obtained only by means of massage. He applied this to the abdomen for fifteen to twenty minutes twice daily, and mild cases required six to eight weeks of this treatment, severe ones three to four months. With it was combined a regulated diet and sometimes the mineral waters of Homburg. Massage has to be proceeded with in the most careful manner in these cases. Sometimes only gentle, firm pressure can be used at first; but after a brief period sensitiveness disappears, stools improve and tympanitis decreases, and then firm, deep working can be employed to act upon indurated connective tissues and adhesions. With the removal of a long-continued tympanitic condition the disturbances dependent upon it cease, and the patient becomes cheerful, sleeps better, breathes easier, and the heart when disturbed resumes its normal action. Exercise must be indulged in with great caution, and dancing, swimming and mountain-climbing must not be attempted for a long time, for the right ileo-cœcal region in the majority of cases is apt to remain a *locus minoris resistentiæ*.

A military surgeon suffered from an attack of peritonitis (*peritonite stercorale*) in 1875, and again had two similar attacks in 1877. After the last, pain was continuous and aggravated by motion, and there was a disagreeable sensation of heaviness, which was accounted for by a hard tumor of the size of a goose's egg, in the ileo-cœcal region. This itself was but slightly painful to pressure. There was no constipation. The case was one of old inflammatory exudation from the peritoneum when first seen by Dr. Weissenberg, three years after the last attack. The treatment had been by drinking mineral salts, and by the local application of the same. Dr. Weissenberg made energetic use of massage, and then amelioration became manifest and was continuous, and all the symptoms terminated in cure, so the surgeon did not have to leave the service as he feared he would.[1] This case illustrates the increased capacity for resorption of the peritoneum under the stimulus of massage, and shows that this agent exerts a similar influ-

---

[1] Berlin Klin. Wochen., No. 17, 1880.

ence upon indurations and exudations wherever they can be reached by it, whether around uterus, kidney, intestine, or other places. For enlargement of the spleen, kneading with the hand is a very old remedy, especially in the West Indies. But we have yet to learn how much benefit results from it.

Taxis for the reduction of hernia is a sort of external massage requiring skill, tact, and care ; and the dilatation of strictures is an internal massage which can be improved on by combining with it external massage where the stricture is accessible to this. In stricture of the urethra each of these methods has had its advocates ; thus Bardinet made use of internal massage alone by the repeated introduction and withdrawal of a sound, while Prof. Antal employed external massage alone with success in impermeable strictures with periurethral indurations. Prof. Antal gave daily massage from eight to ten minutes and good results were gained in from three to eight days, the callous tissue disappearing and the constricted urethra admitting of the passage of larger sounds. Internal massage acts only upon the thin layer of tissue immediately surrounding the urethra ; external massage causes absorption of the entire hyperplasia, and might often be used in preference to urethrotomy. The external method employed by Prof. Antal is of special value in those cases in which the urethra will not admit of the passage of a bougie. Massage of the pendulous portion of the urethra presents no difficulties, but that of the membranous and prostatic portion can be done only through the rectum.

Here is a summary of six cases treated by Prof. Antal:

1st. The patient urinated by drops. Six centimetres behind the meatus was an impermeable stricture. Emollient applications having failed, Antal tried massage and after three *séances* was able to pass a small metallic sound, and after seven days could pass an English sound, No. 13, and the callus had entirely disappeared.

2nd. The patient passed his urine through a perineal fistula, due to a stricture situated anteriorly. At its emergence from the perineum the urethra was surrounded for a space of twelve centimetres by callous tissue. After massage for 5 days he was able to pass a small metallic sound, and soon after a No. 13 English sound ; in 15 days the fistula was closed and the callus reduced to 3 cm. in length and was much softer.

3d.   For six weeks the patient had only been able to micturate drop by drop.   In the posterior third of the cavernous portion of the urethra there was a stricture with an annular callus 2 cm. broad.   A No. 1 sound could be passed.   After six days with massage alone a No. 13 was passed.

4th.   Impermeable stricture in the prostatic portion of the urethra with a callus as large as an apricot.   After five days with massage the stricture was markedly dilated and the callus disappeared.

5th.   Painful micturition ; the urethra surrounded by callus tissue from the fossa navicularis to the pubic arch, admitting an English sound No. 2.   After eight days of massage the callus was but one-third of its original size.

6th.   Impassable stricture of the prostatic portion; callus as large as an apricot.   After three days treatment by massage a No. 3 sound could be passed.   (*Centralblatt für die Gesammte Therapie*, July, 1884.)

As soon as a bougie can be introduced this forms a good substratum for massage to be used over the stricture.   I have found this combination to work well.

Dr. Le Ritter, a Dutch surgeon, has completely cured two patients, aged 50 and 70 years, of retention due to enlarged prostate.   The forefinger was placed in the rectum and the prostate moved from right to left and vertically three times each way, and afterwards it was firmly rubbed.   This is disagreeable and cannot long be tolerated at a time.   In both cases a small quantity of blood passed from the urethra caused by the manipulation, for which liquor ferri sesquichloridi was given until a satisfactory result.   In one of these cases twenty *séances* were required, in the other fifteen, to enable the patients to urinate freely.[1]

---

[1] Brit. Med. Jour., Aug. 1st, 1885.

CHAPTER VIII.

THE RELATIVE VALUE OF MASSAGE IN NEURASTHENIA AS MET WITH IN EITHER SEX.

" Leonato.— Indeed, he looks younger than he did, by the loss of a beard.
Don Pedro.— Nay he rubs himself with civet."— *Much Ado About Nothing.*

NEURASTHENIA, the background to the picture of nearly all diseases and injuries, whether of an organic or functional nature, for our present purpose may be regarded from other points of view. First, there is the natural and not unpleasant fatigue, the result of an active and satisfactory day's work, from which we recover by food and sleep. Then comes the fatigue from which we do not recuperate as usual, the fatigue of being overworked, worried, or, in common parlance, "played out." Here rest or change of scene is of the first importance, and may be all that is necessary; but if this is impracticable or without effect, tonics and sedatives may suffice. Before rest, change, or medication had been resorted to, it has been my lot to have tried massage in several cases of this class, where it has "lifted them up out of the rut" and been the means of procuring good sleep with vigor of mind and body, so that they were able to proceed with their duties uninterruptedly and as easily as ever. Thirdly, there are the continually wearied, wakeful, and nervous business or professional men with numerous and varying ailments, who have learned by experience that "the labor they delight in physics pain," and who find more relief in work than in rest. Massage will sometimes put such on a higher plane of existence, and give them a zest for work which they have not derived from any other source. But unfortunately the interest they gain on their stock of vitality in this way is apt

to be used up as fast as it accumulates. Fourthly, there are
the neurasthenics who are simply spoiled children, who have
plenty to live on without work, and usually more; who have
little or no object in life, and who can do what they please
and cannot do what they don't please, who take delight in
telling of all the eminent medical authorities whose care they
have been under without any benefit resulting, and who are
never happier than when they can be regarded as interesting
by trying some method of treatment that is novel to them in
order that they may have the final satisfaction of saying that
it did them no good. I well remember one such who defiantly
said she would like to see any physician who could benefit
her. Serious disease came upon her. She could not nor would
not understand the nature of it, and be treated as other people
and she prematurely lost her life. There is no help for such
but to lose their fortune and be obliged to work. However,
they form a large part of a clientèle of massage.

Fifthly, there are those who, in spite of rest, change, and
medication, have become chronic neurasthenics, the result of
business reverses, overwork, worry, loss of relatives, disappointed
hopes, or as a sequel of some affection that has existed in some
part of the system, but which has recovered or become of secondary importance. If in these the symptoms point most prominently to spinal exhaustion, — myelasthenia, — where exercise
easily tires and aggravates, massage will be of marked benefit as a
tonic and sedative and corrective of morbid sensations. Of less
advantage, but not useless, in cases where the symptoms point
about equally to easy exhaustion of spine and brain. In cases
of cerebral exhaustion, cerebrasthenia as some call it, where as a
rule physical exercise can be freely taken, this will be of greater
benefit than massage. The value of the latter here is almost
nil, and might be dispensed with, unless it be for a luxurious
placebo to fill up time and keep the patient from trying something worse.

The links in the chain of symptoms that point out neurasthenia or nervous exhaustion, so that they could be clearly
understood as indicating deficiency of quantity or impairment in

quality of nerve-force, were best connected by the late Dr. George M. Beard, of New York City. His statements will hardly be appreciated until they are independently thought out from observation, and one's own conclusions compared with his. Take for example the following paragraph, the import of which I have known to be overlooked by three of the most eminent neurologists on this continent, to whom Dr. Beard was a next-door neighbor :

"*In cerebral exhaustion, active muscular exercise in reasonable amount and variety may be allowed and enjoined ; in spinal exhaustion, relative and in some cases absolute rest is demanded, or only passive exercise for a shorter or longer time, as may be, according to the special peculiarities of the individual.*"

It does not help matters much to coin a Greek or Latin name for every symptom connected with neurasthenia ; and with all the care in naming and accounting for symptoms that Dr. Beard has shown, there are, I think, two which have not yet been mentioned, and which are, like the others, only exaggerations or perversions of natural occurrences. First, there is what almost every one has experienced, namely, an inadequate conception of passing time. Even with the same person at different times and under the same circumstances, time may drag along slowly and heavily ; at others, it will seem to pass so quickly that it is impossible to accomplish anything, even if working in the greatest hurry. Haste is generally a sign of weakness. Secondly, the cause of wakefulness, in some cases of neurasthenia, is doubtless due to a languid condition or want of nerve-force in the respiratory centre, so that lessened respiratory movements have to be supplemented by voluntary ones which require the patient to wake, the *besoin de respirer* becoming so urgent that respiration cannot longer go on involuntarily. Even in health excessive fatigue may prevent sleep. The cardio-inhibitory and the vaso-motor centres in the neighborhood of the respiratory centre may be affected by sympathy with the latter, or they may suffer from impaired nutrition, deficient nervous energy, or irregular blood supply, as may other nerve centres. But to return, for the relief of incorrect appreciation of time and frequent repetitions of waking up at night, as well as for the majority of other symptoms of neurasthenia, massage has proved an efficient agent in my hands. In this and other respects its action is similar to the primary and agreeable effects of opium and alcohol in restoring tone to the respiratory centre and vascular system, without, however, the injurious after-effects of these internal remedies. In place of headache, drowsiness, and disordered digestion, after sleep from massage, the patient is refreshed and buoyant in mind and body. Massage does more than this: it will often counteract the disagreeable feelings that result from the necessity of taking opiates or too free indulgence in alcoholic stimuli. I know of at least one physician who considers Mr. Bombast, who pretends to have rubbed all the Royal Highnesses in Europe, a good man to finish off a fellow who has been on the spree, by a rubbing. There are some people who are born neurasthenic, go through life neurasthenic, and die neurasthenic. Some of these never know that they are lacking in nerve-force, while a few of them do find out, from an occasional day of good feelings or lucid intervals, if one may say so, that their customary vigor is far below what it ought to be. If this class could have massage for a long period or all their lives, they would get a great deal more out of what makes life worth living for. There are those who seldom feel any lack of energy so long as they are occupied, but who have a hard struggle to rest and go to sleep; and these are benefited by massage. Massage is often the only remedy for numerous and indescribable unpleasant feelings, and the subjects of these who have experienced the relief it affords crave its application, as they do food and drink when hungry and thirsty. But most physicians know how fickle neurasthenic patients are; for, even while improving, they will often suddenly give up treatment, for no apparent reason except it be that they are afraid of getting well.

In these and other cases where massage seems to be indicated, it may be given daily or every other day, locally or generally. If there be no apparent effect from this treatment at or soon after its application, it is well to repeat it daily until the latent energies of the patient seem to be rousing, as shown by increase of comfort, vigor, and sleep. Then the intervals between the massages may be lengthened, but not to the extent that the effects of the preceding manipulation may have entirely passed away before another is administered. In some patients reaction is slow, and they feel better the day following that on which they have the massage, than they do the rest of the day on which it is given. In such cases, every other day is sufficiently often to manipulate. Massage of the back alone will often relieve fulness of the head and headache, and this repeated may be all that is necessary. Massage of the back and head will more frequently be used, but general massage is the best for the majority. I have more than once defeated the object in view by overdoing massage on starting, when, as the sequel showed, fifteen minutes would have been all that the patient could take with advantage. I have sometimes overdone the matter at the patient's own request for a longer application, though I had warned him beforehand; and, following the advice of an eminent physician, I have sometimes used massage too freely. The argument too often used, that massage can do no harm if it does no good, is a dangerous one. When a man understands one branch of the medical profession well, one of the commonest errors is to suppose that he understands all the rest equally as well, as if our knowledge of massage, like everything else, did not come through experience.

The time of day at which massage should be given is, in some cases, of importance. If a patient is not very weak, as a general rule, I prefer the time of day at which he feels the worst, or just before this, so as, if possible, to tide him over this period, which can sometimes be done. When a patient does not sleep well, the later in the day massage can be done the better. When a patient of a nervous temperament sleeps well, massage should not be administered in the evening, as it is very sure to make him wakeful, and this applies to such as are well in their nervous system, but may require only local massage for a joint or muscular affection. These are so refreshed after massage that they do not feel the need of sleep. Patients may be benefited by massage when they are too weak to travel. It requires a certain quantity of nerve-force to sustain life at it lowest ebb; more than this, to receive massage with benefit; still more, to be able to travel; and more than all, to exercise freely.

The following two cases are of interest as illustrating good effects from massage when the usual immediate and agreeable effects of it were absent on account of continued mental and bodily activity. Mr. K., a vigorous adult, had confined himself closely to business, both late and early, and had not taken his usual vacation the previous summer. For eight months prior to his sending for me, in March, 1883, he had suffered from wakefulness. He went to sleep readily, but woke up after four hours and remained awake the rest of the night. He had as a result a lack of energy and aptitude for work, but kept at it. In eight days I gave him seven massages of back and head in the evening. With these, continued and refreshing sleep returned, and he has been well ever since.

Dr. B., in the prime of life and in good health, had been on the witness-stand for several days undergoing a fire of cross-questioning involving vast interests in one of the most useful inventions of the age. Insomnia followed, and it was impossible for him to take a few days' vacation. In order to please his wife, he submitted to massage three times, every other evening, and sleep and vigor promptly returned. A few days' rest and absence from work would either have cured these cases or have made them susceptible to the pleasant effects of massage. This treatment may be having effects, even if not felt immediately.

Mr. S., thirty-three years of age, inherits a feeble nervous system, has always been rather delicate and suffered from a lame back. School life was tiresome to him, and when sixteen years old he went into a store, and there lifting heavy weights increased his backache. For four or five years he has suffered from a constant pain in the back of his neck, and holding out his overcoat at arm's length increases this; carrying a parcel for a short distance also aggravates it, and he has not been able to drive any for six months. A sudden jar, or turning in bed, if not careful, will increase the pain in the neck. There is constant pain over the sacrum and coccyx, so that there are often times when he cannot sit more than a few minutes. The whole spine and the muscles on each side of it are very sensitive to moderate pressure. Here the muscles are soft and delicate, less so on the rest of his body. Exacerbations of pain sometimes occur when the patient is at rest. For eight years he has not been able to use his eyes for reading for more than a few minutes at a time, as more than this brings on headache. He has consulted oculists, who have told him that the trouble was not in his eyes. Prior to eight years ago he was subject to sick headaches. He can now attend concerts and plays if not exposed to cross lights. At times he has suffered from numbness in the region of the shoulders and neck with dizziness; at others from gastric disturbance without apparent cause, and once he had an attack of hiccup that lasted for forty-eight hours, and ether had to be administered to stop it. He has had all sorts of internal treatment and counter-irritation externally with indifferent results. Drs. Fred. C. Shattuck and E. H. Bradford referred him to me to try massage.

Mindful of how easy it would be to overdo in such a case, I began with fifteen minutes' gentle massage to the back. No discomfort resulted, and the patient slept from 9 P.M. to 6 A.M. without waking, which was unusually well for him. Three days later he had thirty minutes' massage to arms and back, and felt very nice and comfortable afterwards. On the day following, he jumped on a horse-car which was in motion and jarred the back of his neck, but no more than temporary discomfort resulted. From June 6th to 21st the patient had daily massage, and with uniformly good result — the dull pain in the back of the neck, the sensitiveness of the muscles and tenderness on pressure over the spinous processes had all greatly diminished, and the erector spinæ mass of muscles had become firmer, and the patient could take much more exercise. Sojourns at the mountains and seashore have never done this patient any good, and he has sometimes been worse at the time and afterwards.

Mrs. R. C., aged sixty-five years, of large frame and fairly nourished, but with her tissues flabby and deficient in tone, had always been delicate, and attained her growth rapidly when quite young. For over fifteen years before I was called to her she had suffered more than usual from insomnia; from timidity in going out alone;

from being easily fatigued by exercise or conversation, though she could read and write all day when alone; from numbness radiating from the coccyx over the glutei and posterior aspect of the thighs, often so distressing that she could not sit, though there was no spinal irritation ; from hyperæsthesia extending from knees to ankles ; from languid digestion with occasional gastralgia and at times looseness of the bowels alternating with constipation; from relaxation of the sphincter ani with prolapse of the rectum ; from too frequent desire to urinate, and also from frequent attacks of palpitation and at times irregular action of the heart not accompanied by murmurs. This lady's grandparents had suffered from gout, the only relatives who had ; and many years ago she herself had suffered from stiffness and enlargement of the finger-joints so that she could not shut her hands, but she got entirely rid of this trouble by manipulation. Her large-toe joints were still sensitive and easily hurt when I was called to her in September, 1883. She then had an occasional good night's sleep, but poor ones were the rule. She would go to sleep on retiring, but soon wake up, and then it was difficult for her to get any more sleep that night. On manipulating the head, I felt that the tissues were more rigid on the left than on the right side, and on inquiry I was told that when she lay awake for several hours she became blind in the left eye, but five minutes of sleep always restored the sight.

This patient had a course of massage for three months, at first daily and later every other day. Improvement was apparent from the first visit, and the final result was very satisfactory, but, of course, complete recovery could not be expected. At first the massage was given for a number of times in the morning, and the first application relieved uncomfortable feelings in the back of the neck and soreness and tenderness of the scalp. At the end of a month, the distressing numbness that radiated from the coccyx had decreased; in five weeks she would start off alone and walk half a mile without fear, and two days later she had great comfort in walking, as the sphincter ani kept well contracted, which it did not before. In two months and five days, the bowels had become quite regular under massage of the abdomen, digestion improved, and flatulent distention had disappeared, and urination was less frequent and less urgent. Before using massage on the abdomen, lime-water and a tonic containing nux vomica were tried, but did not seem to have the desired result. At the end of this time, also, she could take a walk of two miles alone, and had entirely lost that timidity which made her afraid to go anywhere unaccompanied. Her skin and muscles were warmer and suppler, and she had gained in flesh. At this time the numbness over the hips and thighs made a sudden and striking disappearance after one very vigorous treatment of kneading, pinching, and percussing. The hyperæsthesia below the knees remained unchanged. After each massage, the previous weak and irregular action of the heart became strong, steady and regular. In our tonic we tried digitalis, and later arsenic, and the heart responded promptly to each, but its increased vigor caused by these was distressing to the patient, who was conscious of its every beat, so that they had to be abandoned. These invigorated the heart's action so that it was out of harmony with the lack of tone in other organs. Two months and a half after massage was discontinued, this patient still held her improvement, notwithstanding an occasional wakeful night and at times a slight increase of numbness. Two or three times since, when apparently she had relapsed, the relief from a few applications of massage brought her back to her much improved condition. This brings me to say that when troubles which have been benefited by massage return, they are much more susceptible to its influence upon its renewal.

This patient never had any children nor suffered from uterine affection. She had no troubles of any kind external to herself, but at times before I attended her the distress of mind was so great that she thinks she would have gone insane but for the comfort and support of religion. Massage had to be varied in quality, quantity, and time of application to meet the requirements of this case. For a while it did well in the morning, then its effect would become lost, and we would change the time to the evening with good result. Manipulation of the head and back was most usually employed, and sometimes percussion of these regions was added, and these procedures were varied with manipulation, passive and resistive movements of arms and legs.

A case at present under my observation is that of a lady whose only two bright little children died two years ago. She has since been in great distress of mind and prostrated in body, and her sleep has been long, heavy, and not refreshing. She has had general massage every other day for two months, and for the past month her sleep has been more brief and natural, she exercises more out of doors, is in better spirits, and takes food with a relish.

Much of the impenetrable mystery that has long surrounded the nature and treatment of insanity passes away when viewed from the sensible standpoint of Dr. Edward Cowles, Superintendent of the McLean Asylum. He writes me that " in many cases of insanity the depression, melancholy, etc., are but the outcome of neurasthenic conditions indicating the need of rest and improved nutrition as thoroughly as in persons not insane. I think massage is of great value in the treatment of the insane, and the indications for it are the same as in ordinary cases of neurasthenia, except that mental conditions sometimes modify or forbid its use." The law of progress is from the general to the special, and to doctors Cowles and Page more than to any others are we indebted for the valuable information, that rest and seclusion are more likely to aggravate than to benefit cases of mental depression. Dr. W. S. Playfair, of London, suspected that such was the case after a short trial of these means, and so expressed himself in the *Lancet* of Dec., 1881 ; but sufficient confirmatory evidence was wanting until Dr. Cowles's report for 1882 appeared, and the same testimony was again given in Dr. Page's report for 1883.

Symptoms akin to those found in locomotor ataxia may be got rid of by means of massage, as the following case tends to show: Mr. P. H., 46 years of age, of slight, wiry frame, active and enduring, for several months before I was called to him had suffered from weakness, numbness, feelings of constriction and inco-ordination of his legs, and was unable to stand on one leg and put on his sock as formerly. These symptoms all disappeared after seven massages in three weeks, following which he took two weeks' vacation, and after that he continued well for a year, walking to and from his place of business, a mile and a half each way, besides being about on his feet nearly all the time. At the end of a year, the same symptoms returned, but to a less degree, and they were accompanied with sweating of the legs to an unusual amount. His family physician gave him atropia, which he only took for a short time. I gave the legs massage eighteen times in seven weeks, and the muscles gained in size, tone, and firmness. He has since, now twenty months, continued well and active, and is on his legs all day.

In this case, control over the bladder and tendon reflex were normal, but there was a little unusual urgency when called to empty the rectum.

The following four cases were kindly referred to me by Dr. S. Weir Mitchell. The first was a clergyman, thirty-five years of age, who when a child had frequent attacks of rheumatism. His duties as a pastor had been arduous, but he was not aware of unusual exertion or fatigue until he broke down, five years before I saw him. The neurasthenic symptoms were about equally divided between cerebration and motion, and developed in two or three days. Rest, travel, and tonics had been tried without result. When called to him, he could take short walks to the extent of three miles a day, and slight efforts at using his brain fatigued him. He then weighed one hundred and sixty pounds, fifteen more than his usual weight. The treatment prescribed was milk and cod-liver oil in abundance, a tonic of iron and strychnia, an hour and a half of absolute rest three times daily, and general massage once a day. Massage made him luxuriously and agreeably tired, so that he slept like a baby afterward, and this ought to have been beneficial, for he suffered from dull headache with uneasy mental and bodily tension, restless nights, and spinal irritation. This treatment was kept up for five weeks and a half, and the patient gained twelve and a half pounds, and though he slept better and spinal tenderness had disappeared, yet he was more easily tired at the end of this time than he was at the beginning of it. During this time the correction of astigmatism was not of such marked advantage to him as would have been supposed. Once while I was

attending him he suffered from more than usual headache for a
week, which disappeared suddenly coincident with an attack of
muscular rheumatism in the shoulder. Both were temporarily
relieved by massage. In twelve days after massage was discontinued, the patient was much stronger. This case received but
little benefit from massage, and perhaps he would have been
better without it.

The next three cases were vigorous young men who could
exercise freely in the open air and walk half a dozen miles at a
stretch, but who had not been able to use their brains in continued study for several years. They had massage from four to
eight weeks without improvement. Indeed, this treatment at
times seemed to make them irritable and hyperæsthetic. In
these cases it is not unlikely that the mischief was kept up by
the injurious effects of long-continued overfilling of the cerebral
vessels, resulting in their enlargement and loss of contractility.
These changes once developed, it is difficult to overcome them.
Muscular exercise and cold to the head would be indicated.

A gentleman about forty years of age, strong in his arms and
body, but always so weak in his legs that he could not walk
more than a square, was sent to me for a course of massage,
half-hour applications to be given him. Five *séances* in three
weeks aggravated the weakness of his legs and used him up
generally. If the length of the massage had been left to me, I
should probably have begun with fifteen minutes in all. In
1870, a vigorous man in the incipient stages of locomotor ataxia
was sent to me for massage. He was given daily massage for
three-quarters of an hour, and at the end of a week he visited
his physician, who, without waiting to question him, at once pronounced his walking better. "Yes," said the patient, "I have
had three times as much massage as you to told me to have!"
Not long after this, this same physician was prescribing massage
by the hour, and soon many other physicians were following
his example.

For the weakness and neurasthenia attendant upon *diabetes
mellitus* massage and exercise short of fatigue have, of late, been
highly recommended for no other reason than that the patient is

benefited. Activity lessens the amount of glycogen in the muscles. The transformation of glycogen may be one source of muscular power. Diabetes occurs most frequently in those who lead sedentary lives, and it is often associated with hyperæmia of the liver and kidneys. Exercise and massage would therefore be rationally indicated to make more blood go through the muscles and external tissues, so as to relieve the congestion of these internal organs. But, of course, attention to diet would be of primary importance.

Prof. Finkler has tried general muscle-kneading in fourteen cases of diabetes. The patients were at first *masséed* daily, and later, morning and evening, for twenty minutes, over all the muscles of their bodies. The diet at the same time was mixed. A few of the patients were confined to bed, others were able to go about, and some did severe manual labor. The result on the whole was favorable, as shown by decrease in the quantity of urine and of the sugar contained in the same, diminution of thirst, return of perspiration, and increase of body weight. After three months' treatment of one patient the sugar entirely disappeared from the urine, and this remained absent for three months after the last massage, then the sugar reappeared.[1]

Zimmer has pointed out that well-developed muscles, even when at rest, are capable of disposing of much more sugar than feeble muscles. As diabetes often occurs in fat people with feeble muscles who cannot exercise, massage, in such cases would be all the more necessary.

In this connection it seems of great significance that a temporary diabetic condition may be induced by any thing which hastens the circulation through the liver, or increases its supply of blood. Observers have met with this result from a variety of causes, but that which particularly interests us is that Bernard found that in dogs the venous blood may have traces of glucose after the abdomen has been subjected to severe pressure or manipulation over the region of the liver and after continued struggles or convulsive action by which the abdominal organs have been forcibly compressed. A saccharine

---

[1] Schmidt's Jahrbücher, Bd. 213, p. 218.

condition of the urine has also been observed in man after a bruise in the right hypochondriac region. In various animals Schiff found that compression of the abdominal aorta for ten minutes, or tying the principal blood-vessels of one limb, might induce a temporary condition of diabetes by accelerating the hepatic circulation.

Dr. T. Lander Branton had a patient who at one time was a tall, powerful man, of active habits in the open air. Some time previously he had suffered from asthma, which had left him, and he became liable to attacks of pain and vomiting. The case was thought to be one of neurotic dyspepsia, and for two years he became more and more emaciated, until he had the appearance of a living skeleton. Only once before had Dr. Branton seen a man so thin, and that was at a show. Under massage and forced feeding his muscles enlarged so that he might have joined a Highland regiment and worn a kilt without being ashamed. From being a simple skeleton he became a well-developed man.

CHAPTER IX.

LOCAL MASSAGE FOR LOCAL NEURASTHENIA.

" I fear too much rubbing : good night my good owl."—*Love's Labor Lost.*

NEURASTHENIA, as I understand it, may be either general or local, affecting the nerves or nerve-cells of all, or any part of the cerebro-spinal or sympathetic system. Its manifestations are those of exhaustion or too easy exhaustibility of nerve-force ; and its pathology, malnutrition of the nerve-cells involved, with concomitant instability of their circulation in the form of anæmia or hyperæmia, or alternations of these. It predisposes to, it accompanies, it results from disease ; the nervous shock and the tedious recovery from injuries point to other sources, and it may.be caused by overwork, worry, or sheer laziness. The agreeable fatigue after a satisfactory day's work that insures sound sleep may be regarded as a healthy form of neurasthenia, if the Hibernianism may be pardoned.

It is a matter of common observation that those who are compelled to hard manual labor seldom suffer from nervous prostration ; and amongst the more fortunate who may be predisposed to neurasthenia, those who are deeply interested in some hobby or occupation that keeps. mind and body active, have found the best means of prophylaxis. The same 'means that serves for its prevention also supplies us with a clew to one of the most valuable agents that can be employed for its relief or recovery. Exercise keeps the circulation active, but requires effort of brain, spinal cord and nerves, as well as muscles, at a time when our object may be to afford rest to one or all of these parts of an overtaxed nervous system. Massage supplies this want, and will keep the circulation going with a minimum or no expenditure of nerve-

force from the patient; and deep massage without friction will
lessen the beats of the heart, and afford it rest also.   Nay more,
for it is getting to be the fashion not only amongst the laity, but
also with some physicians, to say that massage imparts energy to
the patient, though I confess I do not exactly understand what
this means.   Certainly, many who submit to massage feel much
more vigorous, light, and supple after even the first application
than they did before it.   But may not this rather be owing to the
rousing of their latent energies, and restoring the equilibrium of
their forces, by facilitating the circulation of blood and lymph,
and the transmission of nerve-force ?

I have previously stated elsewhere that in cerebral exhaustion
the relative value of massage was almost nil, and that out-of-door
exercise was of paramount importance; but I have since found
reason to modify this in favor of more massage and less exercise.
In such cases, massage of the head alone daily, or every other day,
is better than applying it all over the patient, unless there be a
rare idiosyncrasy that will not allow the head to be manipulated.

There are people, not a few, who, when using their brains, suffer
from uneasy sensations in the lumbar or dorsal region, and these
discomforts continue after the cessation of study, causing wakefulness.   Generally, there is also some spinal irritation in the region affected.   In such cases, massage of the back alone will often
induce sound sleep, and, next day, the patient feels inspired with
faith, hope, and courage, in place of doubt, dread, and fear of meeting appointments.   With these cases, a much more marked effect
is produced by local than by general massage, except when the tenderness of the muscles and spinal irritation is extreme, unfitting
them for every kind work, and then the massage should be general, omitting the back at the first *séances*, but gradually approaching it at subsequent ones.

In other cases of what may be called local neurasthenia, if the
term can be allowed for this purpose, such as writer's cramp, or
the cold, small, and feeble muscles resulting from injury, disease,
or disuse, massage and exercise, carefully adapted, have given excellent results.   To these have recently been added another affection ; namely, laryngeal cramp of musicians and speakers, for the

local treatment of which electricity and massage are considered the most effectual measures.

It is not the purpose of this paper to go into the' details of applying massage, nor to consider its minute effects; but I think it will be a revelation to many to experience either in their own heads, or to observe in those of their patients, the light, comfortable, delightful feelings that are produced by the resistance of a skilled manipulator to forward, backward, and lateral movements of the head. The impression is that the interior of the head has been benefited, and the effect is hardly secondary to massage, which rather gives the impression that the exterior has been improved.

The following cases seem to me sufficiently worthy of notice as examples of the conditions mentioned:

CASE I. A. J., twenty-three years of age. Three years prior to my being called to him, he had been winning races at college at the same time that the functions of his brain flagged, and study had become so irksome, producing headache and insomnia, that he gave it up for a year. At the end of that time he returned to college for a year, and, to use his own words, "patched up and graduated," and for the year before I saw him he had been trying to recuperate by resting at home. At this time, even walking sometimes produced discomfort in his head. At my first visit he had been suffering from headache, with tolerably acute pains in the external branches of the fifth pair of nerves, and had had but little sleep for four nights. The immediate cause of this had been too much conversation with friends on the evening of a holiday. Massage of twenty minutes to the head alone, in the evening, almost completely relieved the headache and neuralgic pains, and was followed by an excellent night's sleep. After this, massage of the head with resistive movements to the muscles of the neck, was repeated seventeen times in twenty-four days, and the improvement in sleep, in comfort of the head, and in the power of using his mental faculties was so great, that it became a serious question whether he should not abandon a six months' sea-voyage that he had engaged. Marks of improvement that may be mentioned were: that when he had an occasional wakeful night, he felt no

worse on the following day ; he had none of his former anxiety
in taking charge of his class in Sunday-school; he attended a
large party late one night without any after-effects, and he walked
about freely, and all while he was preparing for an absence from
home of six months or a year. Medicine had been laid aside before massage was tried in this case.

CASE II. Rev. D. L., aged sixty-six years, has a good appetite
and is well nourished, weighing about one hundred and eighty
pounds. For twelve years he had suffered much from wakefulness.
He required from eight to nine hours of sleep, but seldom got more
than five or six hours of broken, unrefreshing slumber. At times
he would fall asleep soon after retiring, to wake up in a short time;
at others he would lie awake for hours before getting to sleep.
Besides discomfort about the head, he had still more distressing
dull aches and uneasy sensations in the lumbar region, aggravated
by study or wakefulness.

He found some relief from giving up his ministerial duties ten
years before I saw him. He came to me on the 25th of January
of this year, and, after thirty-five minutes of massage on his head
and back at noontime, he passed the remainder of the day in comfort, and that night and the following had seven hours of sleep each,
so that when he came to me on the second day after massage he
was hopeful and radiant. Massage was repeated at 11 A. M. on
head and back, with increase of comfort to the patient. He did
not sleep so well the following night as he did the two preceding
nights, but he realized that he was quiet and serene, and felt that
he was resting, and next day was refreshed. This day he had a
refreshing sleep of an hour and a half in the afternoon, which he
never could obtain when well, and that night slept steadily for
seven or eight hours. This patient had massage three times weekly, at or near noon, for seven or eight weeks, and the result of the
first week is a fair average of the succeeding weeks : five good nights
of sleep out of six, with a nap of an hour or two in the afternoon ;
and, when wakeful, he felt that he was resting, and, the following
day, was not miserable from loss of sleep, as before massage ; vigor
of body and mind gradually increased, and he could take part in
lectures, sociables and other evening entertainments without loss

of sleep, as formerly. Mild tonics and stimulants always made this patient worse. An epiphora that had troubled him for many years disappeared under massage of the eyelids.

CASE III. Mr. E. B., thirty-five years of age, had been in good health for several years, and attended to his business, which involved great detail, from 9 A. M. to 6 P. M., with an hour off at noon for lunch. He had remained in the city all the previous summer, and felt very well when he went away for a vacation of several weeks to Colorado, returning on the 20th of November. From the time of his return he began to suffer from headache, which caused him to be out of his office several hours daily; and, by the end of three weeks, this became suddenly so much worse that he was obliged to leave his office altogether. There was a slight elevation of temperature, but still he had a good appetite, and slept well. I gave him massage of the head on three successive days, and the headache was relieved only while the head was being manipulated, and for a short time afterwards. When massage was being done on the right side of the head, the ache would disappear and increase on the left side; and, on doing both sides they were relieved, and the ache increased in the back of the head; and on *masséing* the back of the head, the discomfort would disappear from there and increase in the forehead, and, on manipulating this region, it would disappear altogether for a time. In other cases, I have chased pain in this way all over the patients without being able to dislodge it completely, only temporary relief being afforded at the place of application.

CASE IV. will serve to show still farther that it is not always well for those who are inclined to nervous exhaustion to give up their employment when working easily, and go away on a vacation. Miss M. P., thirty-four years of age, teacher in a High School, had been subject to headaches all her life. Her parents had highly nervous temperaments. One year before I first saw her, the headache had been so severe that she was confined to bed for seven days with pains all over her and elevated temperature, and since then the headaches have been more frequent and more severe than before, and usually accompanied with nausea and vomiting. Evidently the case was one of migraine. The ache was on the left side of the head, in the left eye, and more especially over the left temporo-parietal region, accompanied with a crawling sensation at the back of the head, and soreness of the muscles of the back of the neck. The left side of the face was smaller than the right. She had great weariness and weakness in her arms, so that it tired her even to raise them. In the cervical and upper dorsal region, there was much tenderness on pressure over the spinous processes. Her appetite was good, bowels regular, and she slept well. Notwithstanding the increase in frequency and severity of headaches for a year, she had gained in weight, mainly adipose, so that she weighed one hundred and sixty pounds, her ordinary weight being one hundred and twenty-three pounds. She continued her duties as a teacher, and found that she felt better when occupied in this way than when not feeling compelled to do anything. Saturdays and Sundays were her poorest days, and a vacation of two months in the South ten months before she came to me was of no apparent benefit to her, but she thought, made it all the harder for her to begin her professional duties again. Correcting examination-papers fatigued her more than anything else. She could walk three or four miles with ease.

The first massage of thirty minutes on the head alone left this region "perfectly comfortable," until the second massage was repeated, two days later, when this comfort was extended to the manipulated regions — head, neck, arms, and shoulders — and a burning sensation between the shoulders was also relieved. In four weeks, a continual wooden, numb sensation of the left side of the head was not only relieved temporarily, but did not return ; and, corresponding objectively to this, her tough, indurated scalp had become soft and supple. Sixteen days after she came to me she could not stand and read on account of weak and uneasy sensations in the back of her neck, as if her head would drop backwards when she attempted to hold the book, but she could sit and read with ease. This was at the catamenial period, when she was generally worse in every way. But, two days later, she was much surprised to find that she had recuperated more quickly, and to a greater extent, than ever before. This patient had massage twenty-five times in ten weeks, with increasing improvement, and this continued after treatment was omitted, so that she was practically well — sufficiently well to enjoy her summer vacation, which helped to confirm the benefit previously received. Six months later, she reported that she had continued quite well. The following spring her troubles returned in the same way, but less severely than formerly. They were speedily removed by massage, and staid away for a year, when, again, there was a slight relapse, and more speedy recovery under massage. At times she had found nux vomica more beneficial than any other internal remedy, but even this had lost its effect before massage was tried.

CASE V. Mr. J. B., aged thirty-three years, has always been a nervous man. He always felt fatigue in the lumbar region, but this he regarded as a matter of course, and he was always capable at business until six years before I first saw him. At that time he was in an elevator which was being tested against sudden falling by means of some " sure patent preventive." The experiment failed, and the elevator fell eighty feet with six men in it. While descending, our patient sat as tailors do, hoping thereby to diminish the shock of stopping. He got out apparently none the worse, walked four squares to his newspaper office, for he was then an editor, and dictated an account of the accident. He staid at home for five or six weeks, but was not confined to bed. After this he resumed his duties, but it was eight months before he could walk a mile. For a long time, in the evening, the region of the spine was painful, but relief was often found by pouring cold water upon it; at other times, from very warm water. During vacation he was perfectly well, and played lawn-tennis. For three weeks before I saw him he had suffered from pain in his back and legs, and could walk but a very short distance at a time. Conversation and reading quickly tired him, and part of either would escape his attention. A few hours at business would cause nausea and headache, and make him feel generally used up. Appetite, bowels, and sleep were in a normal condition.

At my first interview there was much tenderness on pressure

over the spinous processes and muscles of the back ; but, after three massages in nine days, they could be manipulated quite vigorously. After six massages in eighteen days, he was practically as well as ever. Manipulation was exceedingly agreeable to this patient, and, while it was being done on either leg or hip, the agreeable sensation was felt in the back and in the other leg and hip, as well as at the seat of application. About once a year, usually in fall, after his vacation, this patient finds himself used up, as just described, and he has learnt by experience to rely on the prompt relief afforded by massage. He has also found wine of coca of some use. I have frequently made similar cases worse by using massage too vigorously to begin with.

CASE VI. When Mrs. M. W. came to me in October, 1884, she was fifty-eight years of age, and weighed 213½ pounds. Her adipose tissue was supple, and of good consistency. She had then been suffering for three years from a continual distressing feeling of weakness in the right leg and thigh, which first made its appearance when there was some enlargement of the internal saphenous vein, but this had long ago disappeared. On examination, the whole limb seemed normal in every respect. The patient was not at all of a nervous, hysterical, or imaginative temperament, having been at the bombardment of Fort Sumter, once in a steamboat explosion, and once made a long voyage in a vessel with the cargo shifted, so that there was imminent danger of the vessel upsetting, besides having travelled twice round the world, on one of these occasions taking command of a vessel for four weeks amidst shoals and breakers in the China Sea.

A walk of a square was as far as the patient could go with comfort, and a walk of one-fourth of a mile caused great fatigue, and increased the feeling of weakness. She had tried absolute rest for one, two, and three months at a time, during which she lost flesh, but the limb did not improve. At my request, she omitted potatoes, sugar, and butter from her diet, and began walking for two minutes every hour during the day, which was increased daily one minute every hour. Massage was given to the leg, thigh, and hip three times weekly. The first time it comforted and rested the limb, and after this passive and resistive movements were also given, which at first tired the limb, but this was at once counteracted by manipulation. At the end of two weeks she could walk half a mile without fatigue — twice as far as she could before with great fatigue — and a distressing pain that previously came after slight exertion at the exit of the sciatic nerve had not been felt for a week. At the expiration of four weeks the patient walked a mile and a half with ease, feeling but slight general fatigue thereafter, and the limb that had been weak was not so tired as the other. It was by her own wish that massage was repeated occasionally for a few weeks longer, and she has continued well ever since. Under the restricted diet she lost seven and a half pounds, and no doubt but this aided her recovery.

CASE VII. Miss E. H. was thirty-nine years of age when I attended her in the winter of 1883–84. She is irregularly astigmatic, and suffers from headache, and this is worse at the menstrual period, which recurs every three weeks and a half, accompanied with pain. She suffers much at times from indigestion. She is a lady with a strong mind, a clear intellect, an unwearied conversationalist, and, in the language of her physician, who sent her to me, "she is a preëminently hyperæsthetic subject, and would be hysterical, did not the brain govern the *cerebrum abdominale*." For five years she had suffered with pain in her right knee, impairing locomotion, and, the latter part of this time, there was pain also in the outer and posterior aspects of the thigh, where the muscles were considerably atrophied — so much so that her other discomforts seemed small in comparison with those of the limb. The trouble in the limb came when she was run down from nursing a sick relative, and, coincident with this, a severe cough that had been increasing every winter disappeared, and did not return. During these five years under rest, with and without fixed dressings, changes to country and seashore, the use of tonics and sedatives internally, and blisters externally, there would be sometimes a little improvement in the knee, but always followed by speedy relapse on slight or no provocation, such as accidentally hitting it against something, or being obliged to use it a little more than usual. At times, the pain was relieved by walking; at others, made worse. It was aggravated by cold weather and by riding in a carriage.

Examination showed that the affected limb was much smaller than the other, the skin cold and dry, the muscles atrophied, but there was nothing especially noticeable about the knee, save slight puffiness and great tenderness on pressure upon the internal condyle, not in the skin. Owing to pain and weakness, which were aggravated by walking, she could take but a few steps when massage was begun, and the only symptom then in her favor was steady sleep. Massage was applied three times a week for eleven weeks and a half, being omitted for a few days at one time, on account of unusual pain in back, stomach, and intestines. For the first four weeks, massage, with gradually increasing exercises, was confined to the affected limb, with the result that she was, at the end of this time, taking four walks daily, of ten minutes each, besides exercises of standing on tiptoe, stepping up two steps at once, holding the limb out extended, and elevating it sideways when lying down. From the first, the skin became warmer, softer, and suppler, and the muscles fuller, as shown by an increase of one-half inch around the calf; one-fourth inch at the knee; one-eighth inch three inches above the patella; and one-fourth inch seven inches above the patella. When treatment was discontinued, these gains were one-half inch, five-eighths, three-eighths, and seven-sixteenths, respectively. But at the end of four weeks the pain was still about the same, notwithstanding the improvement in locomotion, nor had it entirely disappeared when massage was given up.

As soon as she made known the head and abdominal troubles, massage was applied for ten minutes to each of these regions also, which was during the last seven weeks of treatment. Headache was improved, sleep became more refreshing, digestion easier. During the last eight or ten days of treatment, it became evident that, though the patient was much better, and could go about much more freely on foot and in a carriage, she had come to a stand-still, and consequently the treatment was terminated rather sooner than she wished. A year later I saw her and she was the picture of health. Her appetite was enormous and digestion good. She had gained many pounds in weight and could walk freely, but still suffered from headache. She considered, and I think rightly, that massage had given her a start, and improvement had continued since it was omitted, for no other treatment had been used.

Amongst people who may be considered perfectly well, there are few, if any, who have not some weak points. When fatigued or worried I suffer from tension and dull ache throughout my whole right side. In September, 1884, when in Paris, I had one man give me half-an-hour's massage on my right side only, at 2 P. M., and another half-an-hour's massage on the same side at 5 P. M. The manipulation was slight, superficial and rapid, and at the time of its being done seemed very ineffectual. But that night I never slept so sound in a railroad-train in all my life, as I did from Paris to Calais, and while crossing the channel I was not even sick. Two days later I played deck quoits all one afternoon when the thermometer was 80° in the shade and the ship rolling. Next morning my playmates could scarcely get out of their berths, they were so stiff and tired, and so was I, but the fatigue was all confined to my left side and not to the right as formerly.

It may be said that these were not very sick people, but they are cases that prove troublesome to physicians, and they were certainly in conditions which any one of us would gladly be freed from. It is not necessary that I should dwell here upon extreme cases of nervous prostration that have been treated by absolute rest, forced feeding, massage and electricity. I could give further details of the above-mentioned cases, and also of similar ones, which would seem to justify the following conclusions :

(1) That massage induces sleep.

(2) That even when massage is applied in the forenoon its soporific effects may not disappear before bedtime ; though in general the later in the day massage is used for promoting sleep the better.

(3) Disagreeable feelings of drowsiness and languor do not necessarily intervene between massage in the forenoon and sound sleep at bed-time. Aptitude for rest or work generally follows massage. The mind is clearer, the mental faculties work easier and longer, the muscles are suppler and do not tire so soon.

(4) When people are wakeful after massage they may not be restless nor feel the loss of sleep on the following day.

(5) Spinal irritation is relieved or disappears under massage.

(6) For local neurasthenia there is no need of general massage, unless the whole system be secondarily influenced.

(7) When affections have come to a stand-still under massage, improvement may yet go on after massage has been discontinued.

(8) For improving the nutrition of nerves and muscles, restoring natural sensation and motion, massage may succeed when other means have failed.

(9) Deep massage without friction has proved of more value in my hands than all other forms of massage put together, in the cases herein considered.

(10) Massage can be overdone, producing opposite effects from a moderate application.

(11) Besides massage, carefully-graduated exercises at regular times, are valuable accessories in the restoration of motion.

(12) Massage is not the only means of treatment for neurasthenia. Its selection is usually decided upon after the failure or exhaustion of every other means; in the same manner that the shrewd old divine decided that it was not wise to let the devil have all the good tunes to himself.

CHAPTER X

MASSAGE IN AFFECTIONS OF THE CENTRAL NERVOUS SYSTEM.

When minds are joyful, then we look around,
And what is seen is all on fairy ground ;
Again they sicken, and on every view
Cast their own dull and melancholy hue.—*Crabbe.*

Ross, in his " Treatise on Diseases of the Nervous System,"
says that massage often succeeds in organic and functional
paralysis. It certainly does not often succeed in organic paralysis. The benefits that may result from massage or any other
remedial measure in disturbances arising from organic changes
in the central nervous system, or in any part of the body, will
depend more on the nature of these changes than on the merits
of the treatment, however appropriately and skillfully it may be
employed. So many variations are seen in the course of paralysis of organic origin that the influence of massage in modifying
these is difficult to determine, even if it were judicious to make
use of this treatment from the commencement. When paralysis of central origin has come on suddenly, I prefer to abstain
from the use of massage until the perturbation in general has
subsided, and the patient has become somewhat accustomed to
his unnatural condition. But, in the mean time, while thus
waiting to spare the nerve-centres any supposed extra-commo-
tion, the peripheral pathological changes are gaining ground
which later may only be imperfectly overcome.[1] These are,
interference with the supply and return of the circulation owing
to the accelerating influence of muscular contraction and relax-

---

[1] Dr. S. G. Webber is of the opinion that in cases of cerebral hemorrhage, a few weeks
after the attack, massage can be used with benefit to the nutrition of the muscles; and
later, after five or six weeks, electricity may be given (Treatise on Nervous Diseases).

ation being absent or diminished ; and, as a result of this, variation of temperature, usually lowering, and passive hyperæmia
or ischæmia; hypertrophy of interstitial connective tissue with,
at time subsequent cicatrical retraction, giving rise to contractures and atrophy of the muscular fibres ; formation of adipose
tissue or fatty degeneration ; in a word, vaso-motor and trophic
disturbances. These are all rational indications for the use of
massage, either as a preventive of these changes or as a palliative of them when they have taken place ; but if the nerve-
centres are impaired beyond recovery, or secondary pathological changes have occurred, the prospect of benefit cannot be encouraging. My experience of massage in a number of cases of
paralysis may be briefly stated by saying that, in the absence of
severe pain, obstinate contracture, or tonic spasms, this agent
has proved useful in improving the circulation, temperature,
and comfort of the parts affected. When, in paralysis of spinal
or cerebral origin, recovery follows under treatment, we must
conclude that the central disturbance had entirely passed away,
and that the force of habit was the main factor that continued
the external manifestations of inaction. But even here, when
the causative conditions have ceased, paralyzed muscles will
not at once resume their former natural condition. Massage,
passive and resistive movements restore them to a sense of existence, enable them to recognize the power they still possess,
and educate this to a higher degree ; and, at the same time, such
treatment affords the manipulator the only means of judging of
the capabilities of the patient and of telling him how to use
them. Sometimes the patient will make better motion against
resistance than without it. This seems to give a sense of support and consciousness of power. Massage, if used early in
these cases, would diminish the evils of inactivity upon the circulation and nutrition, and keep the muscles in a state of readi-
ness for voluntary contraction. When there is partial impairment of motion only, will massage be likely to lead to recovery.

Mr. L., fifty-eight years of age, had been a vigorous, healthy
man. He had been much worried with reverses in business.
While at breakfast one morning, he had an uncomfortable sen-

sation in his head, with slight loss of motion in the left arm, leg, and side, which gradually increased for two or three days. He kept in bed for four weeks, and for six weeks afterwards the affected parts were quite helpless, and his face was drawn to one side — the left. Improvement was gradual, and at the end of a year, when he came to me, there was a lack of control over the arm and leg, with stiffness and awkwardness in using them; but if he slipped on the sidewalk he could use either with alacrity to regain his balance. This patient had massage nine times in three weeks, and the result was that he got rid of uneasy feelings in his head, his power of endurance and freedom of motion greatly increased, his digestion, which was previously feeble, became strong and he looked more robust, bowels became regular, and urgent desire to urinate disappeared. The arm and leg could be used almost naturally. At first he felt remarkably well after the massage, as if moderately stimulated. Later he experienced an agreeable languor from manipulation, and he thought he was being too much mesmerized.

From the Out-Patient Department for Nervous Diseases at the Mass. General Hospital, Dr. J. J. Putnam sent me Willie P., five years of age, who had an attack of infantile paralysis (*poliomyelitis anterior acuta*) when ten months of age. But for the diagnosis of an expert, I should have thought the case had been one of hemiplegia, as the history pointed to loss of power in the right side, arm, and leg, the movements of which were still moderately impaired when he came to me, and the nutrition of these parts was little below that of the other side, which was remarkably good. When massage was begun, he could not elevate the arm, though there was an attempt on the part of the deltoid which may have accomplished one-third of the necessary power. In five weeks, he could elevate the arm freely and naturally, and all the other motions had correspondingly increased. Manipulation, percussion, and assistive movements were given for twenty minutes every other day.

One seldom meets with cases so favorable for treatment by massage as these two were. The average of improvement falls far below this, but, where there is any motion at all left, it is likely to be increased by massage and movements.

E. B., aged seventeen years, had, when two and one-half years of age, an attack of infantile paralysis which completely deprived him of the use of his legs then and since, the only motion left being a feeble effort on the part of the right psoas magnus, and iliacus internus, and a moderately persistent contraction of the peronei muscles of the left leg. All the other muscles of the lower limbs were completely atrophied, and his legs were much smaller than his arms ; at times cold, moist, and livid ; at others, warmer and redder than natural. He walked leaning forward on two canes, supported by apparatus on the legs which were projected forward by a side-swing of the body. He had massage every other day for six weeks, and during the last three weeks the legs maintained a more natural and uniform temperature and color, the skin was smoother and softer, and the deficiency of strength in the psoas magnus and iliacus internus had increased so that he could flex the thigh alone when the leg was raised, but not against resistance, and the left foot allowed itself to be held in a natural position, while the patient voluntarily contracted the peronei, any attempt at which at first only brought on perverse spasmodic contraction, with eversion of the foot. Four inches above the middle of the internal malleolus, there was a gain of one-fourth of an inch in circumference of the left leg, showing improved nutrition of the muscles in which there was motion, while eight inches above the lower edge of the internal condyles the thighs had lost one-fourth of an inch each in circumference, owing to the absorption of fattily-degenerated tissue, and perhaps to a less extent of proliferated connective tissue.

In such cases as the last, a warm bath is a good preparation for the massage ; then friction and deep manipulation should follow ; and when there is no motion in the parts, passive motion should be freely given ; when there is slight motion left, but not enough to complete a movement, assistive movements will come into play ; when there is returning motion and more than is necessary for simply moving the part to which the affected muscles are attached, resistive motion within their strength will be used ; percussion when there is a languid state of sensation, motion, and circulation, and a sufficient substratum of muscles to strike upon, but when there is passive hyperæmia, the less percussion is used the better. We often see cases where one group of muscles is paralyzed and atrophied, granting no response to the will, nor to the faradic nor slowly interrupted galvanic current, as, for instance, the anterior tibio-fibular group. Cultivating the extensibility of the opponent group with massage and passive flexion of the foot, together with resistive movements of the whole leg and thigh, made by opposing extension of the leg and thigh, with the opposing force at the ball of the foot, will here be productive of benefit; for in this manner the posterior tibio-fibular muscles will be relieved from their continual state of contraction or retraction, the weak and elongated muscles will be shortened, and both groups will be simultaneously innervated. Medical gymnastics for weak, paretic or paralyzed muscles are based on the fact that exercise of intact muscles stimulates innervation and nutrition of neighboring impaired muscles, and skill in directing these efforts consists in finding out what patients can do, and contriving means for their performance. Massage will probably prove more serviceable in the prevention than in the cure of contractures, stiffness, and anchylosis, whether of central or peripheral origin. In conjunction with elastic muscles which supplement the loss of power in paralyzed muscles, to aid mechanical contrivances to overcome contractures, and to make tissues more amenable to restraint, massage proves useful. After section of muscles, when repair has sufficiently progressed, massage is used to advantage for the restoration of mobility. A famous French surgeon, Malgaigne, I think, has styled massage " the soul of orthopædic surgery."

Cases of locomotor ataxia are benefited by a course of massage from time to time. There is early, frequently after the first massage, improvement in the tone of the muscles, and later, disturbances of sensibility, anæsthesia, and paræsthesia disappear, and patients are apt to think that recovery will result. In the case of Mr. P., who is about fifty-seven years of age, tendon reflex is absent, there is inco-ordination in walking, occa-

sional severe attacks of neuralgia, most frequently in his legs,
diminished control over the rectum, but not over the bladder,
though slight cystitis has long existed. For eight weeks prior
to my first seeing this patient, he had suffered from weakness,
pain and burning sensations in the right knee, and after walking a short distance he would be obliged to stop, as the knee
behaved, to use his own words, like an axle turning without
grease until it would go no farther. The muscles of this leg
were smaller and softer than the other, and there were sensations
of numbness and constriction, most felt about the knee. Under
daily massage for two weeks, the disagreeable feelings passed
away, and the muscles improved in firmness and use, so that in
place of walking a square or two he resumed going to and from
his business, a mile each way. This patient and his physician,
Dr. A. H. Nichols, who kindly referred him to me, were firm
believers in medical gymnastics and exercise *versus* the rest
cure. He has kept them up with massage pretty constantly for
the past four years, and occasionally a tonic internally, and the
result so far is that he has not only held his own, but actually
gained in the use of his legs, while a lateral curvature of the
spine has but slightly increased. This patient has sometimes
walked off his lancinating pains. Though naturally rather a
delicate man, he continues at the head of an extensive business,
and by his persistent exercise keeps himself much stronger than
people in his condition usually are. For the next three years
this patient had two or three massages weekly with the occasional use of the faradic current for a few minutes after the
massage. He retained his powers of motion much better than
we had anticipated, and his general health kept wonderfully
good also, notwithstanding an hypertrophied heart and interstitial nephritis. But when on the 3rd of Sept., 1887, he woke
up and found that his right leg had given out and that he could
not walk at all, it was no more than we had been fearing. He
had had no massage for a month before this except some rough
and inefficient handling by his man-servant, and both of these may
have had some influence in precipitating the loss of power. On
examination, the leg from the knee down was found to be cold, the

muscles were inert and inelastic, and the foot offered much resistance to passive motion, which was limited one-half. There was no power in the anterior tibial muscles to flex the foot, and they gave but a feeble response to a strong faradic current. The muscles of the calf were similarly affected, but to a less extent. Under massage and faradization to the leg, with nux vomica internally, the patient lost ground for the first seven days, the anterior tibial muscles became much atrophied and there were fibrillary contractions. Percussion with the finger-tips caused better contractions than did faradization. There was no trouble with the muscles above the knee. The same treatment was continued, and from about the end of the first week there was steady improvement in the growth of the muscles and in the voluntary contraction of them, so that three weeks later when I tied a handkerchief around the ball of the foot and put a spring balance through this the foot pulled in dorsal flexion (by the contraction of the anterior tibials) sixteen pounds, and three days later nineteen pounds; and while they were contracted by the utmost efforts of the patient it required a pull of fifty pounds to extend the foot. At the end of nine weeks he was walking in the house without the aid of crutches and going up and down stairs naturally, and at the end of three months he was walking as well as he had done for several years before the leg gave out. The muscles of this leg had been so trained and cultivated with massage, electricity and gymnastics, that they were really stronger than the other comparatively well leg. Two years have elapsed, and the patient has held his improvement. From the commencement of his spinal symptoms, eleven or twelve years ago, he has practised daily suspension by means of Sayre's apparatus, besides rowing with elastic tubes on a parlor rowing-machine. The occasional cauterization of this patient's back had been omitted for some weeks previous to the loss of power in his leg; and it was not resumed until he could go without crutches, so that if its suspension had anything to do with precipitating his symptoms, its omission had nothing to do with his recovery, unless to aid it.

If space permitted, I could tell of another locomotor ataxic whose anterior tibial muscles of the right leg lost all power of

motion twice within the same year, and each time he regained full
strength and control over them in a few weeks under massage,
movements and faradization.   Success in these cases was doubtless
due to the fact that treatment was begun early, before the nerves
and muscles had time to degenerate to any extent, and it might
lead us to think that there is probably much more significance to
the statement of Dr. Wm. Murrell than we have attached to it,
namely, that massage gives good results in recent cases of infantile
paralysis.

Fuller in his *Medicina Gymnastica*, 1771, says "there is no
reason why many invalids should not be as much stronger than
others who are similarly afflicted, as gymnasts are stronger than
people usually."   The pendulum of fashion is apt to swing to
extremes in medicine as in everything else.   At present the rest
cure (as some call it) prevails, it may not be long before the
exercise cure will be in vogue.[1]   But, however it may be, the
skill of the physician will always be necessary to encourage the
one or restrain the other.   It would be the height of absurdity
to advise exercise when locomotor ataxia or any other malady
had been brought on by bodily fatigue, at least until a sufficient
space of time had been allowed for rest.   There is a principle
underlying rest and exercise.   The phenomena of life alternate
in an active and passive manner, individually and collectively;
sleep alternates with waking, rest with exercise, muscles contract
and relax; walking is a semi-passive means of locomotion by
which we partly project ourselves and partly fall forward.   Invalids who have been wearing themselves out by repeatedly futile
and exhaustive efforts at exercise of mind or body must have a
long pause of rest.   The most careful judgment is required to
properly limit this, to prevent its evils, and to get patients to resume exercise in minimum quantities with suitable intervals of
rest, gradually decreasing while exercise is increasing.   To preserve the harmony of rest and exercise, and to prescribe the one or
the other to suit different conditions, are important and often
difficult problems.

---

[1] The above was written five years ago, and this prediction is now more than fulfilled,
at least in the United States, for men, women and children in all conditions have taken
to exercising as if their lives and salvation depended on this alone.

Zabludowsky, in the *Deut. Med. Zeitung*, No. 3, 1884, reports a case of locomotor ataxia in which there was a history of syphilis. After medical treatment had improved the patient to a certain extent beyond which he did not seem to gain, massage was used and the patient improved in a remarkable manner.   Schreiber has used massage successfully for the removal of distressing anæsthesia of both gluteal regions in a case of locomotor ataxia (*Wiener Med. Presse*, Mch. 6th, 1881).   The case presented well-marked symptoms of the malady; ataxia, lancinating pains, gastralgia, and paralysis of the sixth pair of nerves.   Sensations of temperature and of contact over the hips were so deficient that the patient could not distinguish whether he was seated on a cold stone or on a wooden bench warmed in the sun, nor whether an object was hard or soft, and this state had existed for five months.   Schreiber knew by experience that anæsthesias in the course of sciatica yield to massage.   Although it is generally considered with regard to tabes that a *noli me tangere* course is the best so far as manual intervention is concerned, yet he resolved to try massage in the most careful manner.   He soon found that it caused no unpleasant effects, and he therefore used it vigorously every day for twelve days with final and complete relief to the patient, which still continued three months afterwards.   The patient kept notes of his case.   November 15th, 1880, after the fifth massage he experienced a disagreeable tension in the parts *masséed* which rendered the ascent of stairs difficult.   November 18th, tension has disappeared and the strength of the muscles increased.   19th, slight sensation, could feel where he was seated, whilst before it seemed as if there were a foreign body between him and a seat.   20th, sensation increased and he could tell whether a body was hard or soft; 22d, greater sensibility to contact; 23d, last trace of disagreeable feeling had disappeared.   The massage consisted of stroking, kneading, and percussing for five minutes every day.   Türk is said to have been the first to prove that slight degrees of anæsthesia may be got rid of by rubbing alone.

Before cutting down and laying bare a nerve-trunk, to stretch it for the relief of the pain of locomotor ataxia, or for any purpose whatsoever, massage should be thoroughly tried, as the action of each

method is somewhat similar to the other, releasing the nerve from
the neighboring tissues that compress it, and producing changes
in its structure and circulation, and lessening its irritability, perhaps by over-inciting it. Massage makes repeated mild stretching,
and might succeed when more violent stretching would fail.
Langenbuch makes use of massage in the vicinity of the wound
after the violent stretching by surgical operations. It is not stated
whether he or anyone else had tried massage before resorting to
such extreme measures. He has operated in one hundred cases of
locomotor ataxia, but gives no summary of results. He speaks of
six cases out of sixteen as being cured of their pain and disagreeable feelings. The indications for the surgical operation of
nerve-stretching are not clear, and evidently it must be done for
luck, for Langenbuch says: "I have operated in far advanced
cases which to all appearances were very unfavorable, when the
patient had not left the bed for two or three years, and have seen
the patient get on his feet again; also I have been able to accomplish very little in relatively early cases which had neither
much pain nor any symptom of bladder disturbance."[1]   Hence the
uncertainty of prognosis. Bardeleben, in a resumé of the results
obtained in his experience, disclaims responsibility for the operation. He only operated when requested to do so. He had obtained
no good results, and it was generally without benefit to the patient.
"We can never know whether the nuclei of the affected nerves are
destroyed or remain so as to favor restoration to the fibres." Dr.
G. L. Walton reports four cases of nerve-stretching for affections
of the spinal cord with no beneficial result (*Boston Med. and Surg.
Journal*). Dr. Mortimer Granville has succeeded in relieving the
pains of locomotor ataxia and other affections by means of percussion over the affected nerves, and he thinks this endeavor to bring
about a natural condition ought to be tried before resorting to so
formidable an operation as firmly stretching a nerve, and for
the time being mechanically disorganizing it.

    There seems to be no doubt as to the benefit that locomotor

---

[1] Berlin. Klin. Wochen., March 20th and 27th, also Dr. S. G. Webber's Report in
Boston Med. and Surgical Journal, 1882.
    Virchow and Hirsch's Jahresbericht, 1881.

ataxic patients derive from suspension by the head and axillæ; but how this acts is still largely a matter of conjecture. That the spinal cord receives any stretching at all in this way is very doubtful, though the spinal nerve-roots may receive some. The vertebral column is elongated from 2½ to 4 centimeters (1 in. to 1.6 ins.); its muscles and ligaments are stretched. But this is accompanied with great danger in some cases. The same object can be obtained much more safely and effectually by making extension and counter-extension at both the head and feet simultaneously, with the patient in a horizontal position, or even on an inclined plane. Suspension from the elbows by the sides can act only upon the muscles of the trunk and chest, and in the same proportion take it off the spinal column, if suspension by the head be used at the same time. A better and more efficacious way than any of these is by passive or active flexion and extension of the trunk, preceded and followed by massage. For accomplishing this any physician may have his method called after his name, according to the skill and ingenuity he may display. The experiments of Hegar [1] throw much light on this subject, and are regarded as the most satisfactory. He laid bare the dura mater of the spinal cord in the dorsal region, and inserted into it two bright threads, at a distance of 12.5 centimeters from each other, the cadaver being in a horizontal position, back uppermost. Moderate flexion, produced by placing blocks under the chest and abdomen and bending the neck, the legs being free, made the distance between the threads 13 cm., an increase of 5 millimeters. Strong flexion in the same manner made the distance 13.2 cm., a gain of 7 millimeters in all. This was increased only 1 millimeter more by flexing the thighs upon the abdomen, with the knees extended. The cadaver was then placed flat, as at first, and the distance between the threads became 12.5 cm. as before.

Both sciatic nerves were then laid bare, and a strong pull made upon them, which only increased the distance between the threads 1 millimeter. The spinal column was again flexed, and the distance measured 13.1 cm., an increase of 6 millimeters;

---

[1] Wiener Med. Blätter, 1884, No. 3.

and in this position the sciatic nerves were strongly pulled, and the distance became 13.3 cm., a gain of 8 millimeters in all. The next question to determine was whether the spinal cord itself partook in the extension. The dura mater was opened, and two threads stitched into the substance of the cord at a distance of 15.35 cm. from each other. Upon moderate flexion this became 16.1, and upon strong flexion 16.4 cm. Hence we arrive at the very interesting conclusion that the cord itself allows greater extension than does its dura mater. Division of both sciatic nerves had no influence upon the result.

By means of forcible bloodless stretching of the nerve-trunks Cattani obtained the same results as by bloody stretching: tearing of the axid cylinder, stretching of the medullary sheath, with subsequent degeneration and regeneration.

We have every reason to suppose that stretching and relaxing of nerve-filaments, nerve-trunks, and even the spinal cord itself, are as essential for their proper circulation and nourishment as muscular contraction and relaxation, and for the nutrition and welfare of the muscles. But, of course, even this feature might be carried to excess.

The writer has frequently been asked by thoughtful physicians, " Have you ever cured a case of locomotor ataxia by massage ? " as if they really thought it might be possible to do so. Patients with this disease are generally more hopeful of beneficial results than physicians, and the knowledge, for which we have such good authority as Erb and Ziemssen, that this affection frequently comes to a stand-still and improves, and that in some few instances recovery has taken place, ought to inspire us with more hope and zeal to coöperate with the sufferers. Dr. Mortimer Granville says : " In neurasthenia and even commencing sclerosis of the spinal cord with loss of tendon reflex, the most remarkable effects are produced by applying the *percuteur* (percussion instrument) over the spinous processes of the appropriate vertebræ. In a few cases I have failed, but in others — not a few — the locomotor ataxia has been removed or sensibly ameliorated and the general improvement astonishing." ("Nerve Vibration and Excitation," p. 41.)

From a summary of the most recent and comprehensive views gleaned from the experience of the ablest and most trustworthy observers of *progressive muscular atrophy* by Eulenberg, we learn that the prognosis of this affection is generally unfavorable, but by no means absolutely hopeless, for the treatment of it has many successes to boast of; but in order to gain these "*it is necessary to begin as early as possible and to persevere with untiring patience as long as possible; that "absolutely nothing is to be expected of internal remedies;"* and that *"the only suitable and really trustworthy remedies are electricity and medical gymnastics."* Undoubted successes, he says, have followed the use of suitably localized gymnastics in this disease, and it is easy to say that we possess in active and passive movements an agent of especial efficacy for the interstitial changes within the muscles. In a case that had recently fallen under his observation, the process of massage was said to have brought the disease to a standstill.

The following is the only case of this kind in which I have used massage. Mrs. E. R. was an elderly lady and enjoyed a remarkable degree of health, notwithstanding the fact that she labored under *diabetes mellitus* of several years' duration, which was kept under by restricted diet and frequent sojourns at Carlsbad. For six months before I was called to her in January, 1878, she had difficulty in tying a knot, inserting a pin, turning a key, and also in writing, all of which appeared soon after excessive use of the hand in cutting out garments with scissors for the poor. These symptoms of weakened muscles and lack of control slowly increased in an ascending manner, and group after group continued to become atrophied and powerless. They had progressed to a considerable extent in one arm before the other became similarly affected; later, body and legs were also affected, and for a while speech was greatly impaired before deglutition and respiration shared in the malady. Finally, the patient died, four years and a half after the appearance of the first symptoms. Massage with such movements as the patient was able to make, were kept up for two years; resistive movements and exercise with elastic tubes so long as the patient could; then, when she could not oppose anything, simple active motion, and when this failed, assistive movements; and after no effort could be made, passive motion. There was no indication for friction in this case, but deep manipulation and percussion were freely used. After the first five applications of manipulation and movements, it really seemed as if the patient were cured, as she could do with ease what before was difficult, and this improvement was held for a long time, even though the treatment was interrupted by an attack of bronchitis. If Eulenberg or anybody else had seen this patient before and after these five visits, he would certainly have believed with the patient that the disease had been brought to more than a standstill. Later, as the strength was fading out of the muscles, there was a marked improvement after each massage, so evident that, when there was no longer power to flex the arm before the massage, she could do this afterwards, and the deltoid showed the same behavior in raising the arm as the last flicker of strength was dying out. There is no doubt but massage retarded the progress of the disease in this case. I frequently warned this patient of the gravity of her symptoms, and told her that she ought to consult other physicians; but this only made her all the more determined to adhere to me and to massage, the effects of which were so apparent to her. In one case of suspected incipient progressive muscular atrophy that came under my care, the symptoms disappeared entirely under massage, and have not yet returned, now seven years.

Weir Mitchell says: "It is many years since I first saw massage used by a charlatan in a case of progressive paralysis. The temporary results he obtained were so remarkable that I began soon after to learn what I could of its employment." ("Fat and Blood," p. 51.)

Theory and practice do not quite harmonize in the treatment of locomotor ataxia as compared with that of progressive muscular atrophy. In the former, rest is advocated; in the latter, exercise and massage. It seems hardly reasonable to urge on, by exercise, the degenerating cells of the anterior cornua in progressive muscular atrophy; while one would think that in locomotor ataxia the disease in the posterior columns and vicinity would be favorably acted upon by the derivative influence of increased activity of the anterior portions of the cord called forth by such means as exercise, massage, passive and resistive movements. On the other hand, if the degenerating nerve-cells in the anterior cornua are benefited by functional activity, and the evidence is that they are, ought we not to expect that cultivation of the co-ordinative powers would exercise a like favorable influence upon the co-ordinating tracts presumed to lie within the posterior columns? We are told by Mortimer Granville that, in the case of the blind who become ataxic, the muscular sense, being so highly developed, compensates for the loss of sight, and that it is not easily impaired even by paralytic disease, and therefore the distinctive symptom of ataxia is often wanting until an advanced stage of the disease. Could we have a greater argument than this in favor of massage and the cultivation of movements which improve faulty muscular sense? In one of my cases, I have used walking with the eyes shut, and I believe it has improved co-ordination.

In *pseudo-hypertrophy of the muscles*, which is considered a modified form of progressive muscular atrophy, inasmuch as the first stage of it consists of a chronic irritative process of the interstitial connective tissue affecting secondarily the muscular elements, and hence defined by Friedreich as *à chronic myositis accompanied by interstitial hyperplasia of the connective tissue,* massage and hydro-therapeutics seem to have been of value in some cases. As regards the use of these with localized gymnastics, there are as yet too few observations, but they seem to promise advantage in the initial stage of the disease, though at a later period no success is to be expected, at least in the restoration of the affected muscles (Eulenburg).

In a case of *disseminated sclerosis* of the spinal cord that lingered along for many years under the care of the late Dr. Edward H. Clarke, it was thought that the patient declined more rapidly in the summer months when he did not have massage, than during the rest of the year when he did. During the eighteen months that he was under my observation, massage seemed to preserve the nutrition and power of contraction of the muscles and to lessen the tendency to undue increase of adipose tissue to a very late period of his illness. The patient experienced so much benefit from manipulation that he asked to have it on Sunday, as well as every week-day. His pulse before massage was from 90 to 96; after, 84 to 90. Respirations before massage, 24 to 28 per minute; after, 13 to 24. Dr. Brown-Séquard, who saw him late in his illness, stated that he was no worse than he was one year before. Of course, skilful medication had much to do with sustaining the patient. He could lift four hundred and sixty pounds on a lifting machine, and took this as one of his daily exercises, until his disease was far advanced, and at a time when his walking was like that of a drunken man. It is not likely that Dr. Clarke would have allowed this, had not experience proved the usefulness of it.

Encouraging success has attended the use of massage and gymnastics in *chorea* without much regard to the pathology or causation of the affection. It is generally agreed that the seat of this malady is for the most part in the brain, though the spinal cord and peripheral nerves may, and generally do, share in the disorder which is of such a nature as to weaken the force of the nervous system without destroying it. If the erratic movements were due to local disease in the brain or spinal cord, we would not expect much benefit from massage, or any form of treatment. Rest, massage, and abundance of easily digested food have proved successful in the early or acute stage; in the decline of the malady, when slight irregular movements have still lingered, massage, exercise, and calisthenics have done well. Anæmia, chlorosis, rheumatism, endocarditis, etc., should be met by appropriate remedies. But "there are great difficulties in the way of forming a critical judgment of the activity of remedies in this disease, whose duration is so variable, whose course is always subject to spontaneous remissions, and which so often passes away quickly and easily without any medication" (Von Ziemssen). Nevertheless, the medical treatment of chorea is pronounced very satisfactory, and complete cure is the rule in from two to three months on an average. Massage does better than that, if we may believe the published reports, and they are properly vouched for. In 1847, the staff of the *Hôpital des Enfants* in Paris appointed Napoleon Laisué to use massage and movements in the treatment of chorea, and the result was reported to the Academy of Medicine by M. Blache, one of the physicians to the Hospital. From this report we quote the following: "One hundred and eight (108) cases have been submitted to the treatment by massage. Of these, 100 were in the first attack, at the beginning of the affection and severely afflicted; 8 were on the decline. These were divided again into two categories: 34 cases of medium intensity; 74 in which the agitation was as violent as could be. The 34 cases of the first class were all cured on an average of 28 days with 18 *séances*. Of the 74 more serious cases, 68 cases were cured in 55 days with 31 massages. There remained 6 cases, without success, chronic cases which finally got well in 122 days with 73 *séances*.

"Let us speak of the details. Take a patient lying in a bed in the form of a box with upholstered sides to protect the child from injuring itself by its violent and disorderly movements, the child unable to stand or hold anything in its hands, not even able to speak, in a word with powerless will. While three or four assistants hold the little patient on its back and keep it so for ten or fifteen minutes, the 'professor' *massées* with his whole hands the upper and lower extremities and the front of the chest. The posterior aspect of the patient is similarly dealt with, but principally the muscles of the back. A *séance* of this sort lasts about an hour, and is repeated once in three or four days. Each time an amendment is observed in the disorder, and if the patient is wakeful, as is usually the case, calm sleep follows."[1]

"Conclusions: 1. None of the methods of treatment applied to St. Vitus' dance has given so many cures as massage either alone or with sulphur baths.

"2. Massage can be employed in almost all cases without being interrupted by the contra-indications which present themselves to other methods of treatment.

"3. Cure is more durable than that obtained by sulphur baths, and the sedation shows itself in the first days.

"4. As the disorder declines, the constitution is ameliorated in

---

[1] Moniteur des Hôpitaux, 1er Août, 1954. Laisué Du Massage, p. 27.

a marked manner, and patients are cured not only of the chorea, but also of the anæmia which so often accompanies it.

" The exercises are in no manner dangerous. They are of two kinds : passive when the will has no power over the muscles; active, when they can be done." From the way the report reads one would infer that no medication was used, but M. Blache says that analeptics ought to be employed with the massage. Besides Blache, five other physicians testify to the value of massage in chorea as used by Laisué. Less massage and oftener repeated would have been better.

The London *Lancet* for Aug 5th, 1882, contains an article by Doctors Goodhart and Phillips on the treatment of acute chorea by massage and the free administration or nourishment, with rest in bed. These authors state that no treatment is so satisfactory, but that others may prove useful in selected cases. They are evidently not aware of the fact that massage had been used in chorea before they tried it; at any rate, if they had been, they would doubtless have corroborated their results by quoting those of Laisué. They consider the disorder a nervous habit aggravated by neglect and nerve-exhaustion into an acute disease. Three cases of heart disease with fever which speedily proved fatal were not regarded as suitable for this treatment, and it was not tried at all with them. Twelve cases were treated in the manner spoken of, and the advantages proved to be : that when the massage was carefully performed, flabby and poor muscles became plump and healthy; the various groups being manipulated in an orderly manner, it is inferred that some influence was exerted towards restoring more equable nerve-discharges from the centres which control them and dispelling a disorderly habit by an orderly one ; the supplies were utilized to their utmost and without call upon the diminished capital of brain power. These observers modestly state that the success was not in all cases convincing. If any one will take the trouble to read the details of their cases they will find no reason why they should thus have underestimated their success. Marked improvement was observed in every case in the following particulars : a decided increase in weight; rapid subsidence of all the more violent movements ; the extremities soon became warm; the pulse fell and became more regular; the patients slept soundly after massage.

The temperature, which is usually normal in chorea, was found to fall from 1° to 2° F. after massage. The amount of urea excreted was tested daily, but no increase or decrease could be found to correspond with the increased nitrogenous waste. Systolic murmurs and *bruits* disappeared. Massage was given for fifteen minutes twice daily, much more sensible than the hour doses employed by Laisué. Why the treatment is only advocated for the acute stage of chorea, and not for the chronic does not seem clear, unless it be that seeing such marked benefit from it early, led them to expect too much from massage later, when there was less margin for improvement. The last remnants of any affection are generally the hardest to get rid of. If calisthenics had been added to massage, the minor disordered movements that remained and characterized a few of their cases as chronic would have disappeared more quickly. Only in one case was medicine given, opium and bismuth on account of diarrhœa, while massage was *suspended* for three days. In four of the cases the treatment produced a striking result, and it was believed to be instrumental in saving the children's lives. All the cases were of the most unfavorable kind, with, in some, bad family histories as predisposing causes; most of them could not stand, others could not articulate and suffered from night-terrors, with loss of control over their dejections and inability to feed themselves. The relief from these distressing symptoms was too marked to be regarded as mere coincidence. The treatment produced good effects in all, and in several upon the chronic movements also. It is difficult to refrain from giving all the details of the cases, as they are so full of interest and import. In the first case, all active movements ceased after seventeen days of massage; in the second, after eight days; in the third, after twenty-one days; in the fourth, after seven days; in the fifth, after fourteen days; in the sixth, after twelve days; in the seventh, after fourteen days, and night-terrors disappeared, weight increased, and the patient took food well. This case was called a decided failure. What then must have been their successes? In the eighth case, four years of age, after seven days the patient could feed himself, articulate distinctly, pass dejections consciously, and a "post-systolic brush" had disappeared. In case nine, the patient could not stand on admission, but after fourteen days was allowed to get up, and there was no return of the movements. In case ten, after fourteen days of massage all headache disappeared, movements ceased, and grasping power returned. These were treated at Evelina Hospital. Cases eleven and twelve were treated at Guy's Hospital with similar results.

After these two encouraging reports of the value of massage in chorea, my own experience is scarcely worth mentioning. Nevertheless the following case presents some points of interest not brought out in the above one hundred and twenty cases. It occurred at the City Hospital in the service of my friend, Dr. John G. Blake. Mary Wise, nine years of age, was admitted January 16th, 1874. But little could be learned of her history. She had had rheumatism three weeks before, but for several days there had been no pain. The appetite was good and bowels regular, but the child was anæmic. There was insufficiency of the mitral orifice, as shown by a loud murmur at the apex with the impulse. There were spasmodic movements of the limbs, and the child could not feed herself, nor use her left hand. Bromide of potassium in twenty-grain doses with two grains of iodide of potassium were given three times daily, and also six drops of the tincture of the chloride of iron. Four days later there was more control over the limbs and the patient could feed herself. Twenty-seven days after admission, the irregular movements were still violent, especially of the face, left arm, and left leg. It was then that massage was begun, and medicine and electricity omitted. I gave her twenty to thirty minutes' massage every evening, and she slept soundly after this, better than after bromide. She could not at first do anything like calisthenics or gymnastics, but I directed, whenever the irregular movements occurred, and also between their attacks, she should seize the foot of her bedstead, and pull with all her might; in this way the will soon regained control over the muscles and broke up the disorderly attacks. After one week she could walk from one end of the ward to the other, about one hundred and twenty feet, on a straight line formed by the seam of two adjoining pieces of flooring, and at the end of two weeks she was assisting the nurses in carrying liquid medicines. In this case massage more than took the place of the medicines which had done so well, but the effects of which seemed to have come to a stand-still.

In the sixth century, chorea was cured by dancing for three hours, after which the subjects of it were free for a whole year (Schenk de Graffenberg). Horst, a physician of the seventeenth century, says that persons affected with involuntary movements danced during the day and night till they fell on the ground in a trance. Afterwards they were quiet till the next year, when they again felt an agitation and they returned to the chapel at Drefelshausen near Ulm in order to dance (Roth).

In affections of the nervous system massage may be overdone, badly done, or misapplied. These distinctions could not have been very clear to Dr. Althaus when he called the attention of the profession to the " Risks of Massage " in the *British Medical Journal* of June 23d, 1883. Evidently he regards all sorts of massage alike, without reference to quality or quantity. He says that massage, long the Cinderella of therapeutics, has now become as thoroughly fashionable as homœopathy and mesmerism have been. However applicable to obstinate hysteria, he deprecates the indiscriminate use of massage to all sorts of cases of cerebral and spinal diseases in which loss of motion is a conspicuous symptom. He mentions a few pages at the end of a book on orthopædic surgery by Prof. Busch, of Berlin, as the most recent and sensible treatise on massage, wherein it is recommended in writer's palsy, stammering, hysteria, and muscular paralysis, or paresis after poliomyelitis, without, however, saying much in its favor in the latter condition.[1] The doctor proceeds in these words : " It appears to me that diseases of the brain and spinal cord must, on account of the anatomical situation of these organs, be inaccessible to the influence of massage which can only be applicable to more superficial parts of the body. Many of the most important diseases of these organs are of an inflammatory

---

[1] In the British Medical Journal for May 31st, 1879, Dr. Althaus himself speaks approvingly of the use of percussion in infantile paralysis.

or irritant character, either primarily or secondarily, and massage should not be used for their treatment even if the suffering parts could be reached by it. I will here only allude to many forms of cerebral paralysis from hemorrhage, embolism, or thrombosis, which are followed by sclerosing myelitis of the pyramidal strands, and most forms of primary lateral, posterior, or insular sclerosis of the spinal cord." . . . "In most cases, of lateral and insular sclerosis which are now unfortunately much treated by massage and exercise, rest is indicated rather than active exercise, and overstraining of the enfeebled muscles acts prejudicially on the state of the nervous centres. I have recently seen quite a number of instances in which the central disease had been made palpably worse by procedures of this kind; and in a case of cerebral paralysis which was some time ago under my care, the patient had, after four such sittings, been seized with collapse, which nearly carried him off."

To this Prof. Playfair replied in the *British Medical Journal,* June 30th, 1883. Having been one of the first in England to call attention to the value of massage, he held himself in a manner responsible for its having become as fashionable as mesmerism and homœopathy have been, if such be the fact, of which he has no knowledge. If Dr. Althaus compares these systems with massage, the comparison is most unjust, says Prof. Playfair. "Mesmerism and homœopathy," he continues, "are what we all know them to be; massage is a thoroughly scientific remedy based on good physiology and sound common sense, the value of which, in properly selected cases, no one who has any knowledge of the matter can possibly question; and which doubtless, when improperly applied, is capable of doing much injury, as any other powerful treatment may under similar circumstances." He is quite in accord with Dr. Althaus as to the importance of accurate diagnosis before resorting to its use. But if he should adopt the same method of argument as Dr. Althaus, he might, he says, illustrate the difficulty of accurate diagnosis in doubtful cases of neurasthenia, by several very remarkable instances in which patients have been for years treated as subjects of organic spinal disease by some of the most eminent neurologists, in which the error of diagnosis has been conclusively proved by their rapid and complete recovery under appropriate treatment, of which massage formed an important part. He concludes by saying that the results, in well selected cases, are so remarkable that it is not surprising that it should run the risk of being at times injudiciously used, as Dr. Althaus supposes to have been the case. This is only what happens when any new subject attracts attention.

Though these two giants have crossed swords on the subject of massage, yet the assailant has been spared in his weak and vulnerable points. It will be seen that Dr. Althaus at one and the same time denies and admits the influence of massage upon the nervous system. On account of the situation of the brain and spinal cord, he denies that massage can have any influence upon their diseases. This may sometimes be true, but not on account of the situation of these organs, but by reason of the nature of their diseases. He evidently forgets that all external impressions, whether of heat or cold, of massage or electricity, of a chemical or mechanical nature, do, and must affect the brain and spinal cord, otherwise they could not be perceived. He admits that massage, or more likely some violent handling that has passed for massage, by overstraining weak muscles, acts prejudicially on the affected nerve-centres. But massage can be applied so as not to be overstraining, irritating, or collapsing in its effects ; though collapse may occur at any time in some of the conditions mentioned by Althaus, and especially in such a case as he says it did. How many of such cases have collapsed or died while receiving electricity ? Certainly some have, though I have not heard of them. Is electricity to be blamed for this if properly administered ? Assuredly not, any more than if the physician himself had died of apoplexy while administering it. Laisué was careful to have the physicians of the *Hôpital des Enfants* sign a statement that no untoward event had occurred in his massage and exercises of one hundred and eight cases of chorea. If this remedy is so dangerous as Dr. Althaus makes out, here certainly was a chance, at least a coincidence, for harm to follow in these cases which " are so apt to terminate in paralysis," according to Broadbent.[1] The fact that none of them did terminate in paralysis would be a strong argument for massage as a preventive measure against such a catastrophe. In a case of a different kind, I am sure this calamity was averted for a while by means of massage. The patient was a gentleman sixty-nine years of age, who had enjoyed a life of health and activity, but who for many months prior to my visiting him had been deprived of rest and sleep by extensive business cares and sickness in his family. During the day his mind was at times confused, and slight things annoyed him overmuch. In the evening his face was deeply flushed and he suffered from headache. Apoplexy was feared, as several members of his family had died in this way. The promptness with which sleep returned, headache and flushed countenance disappeared, and clearness of mind with vigor of action were enjoyed on the commencement and continuance of massage, left no room to doubt its benefit, for his family and business cares were still upon him during his treatment, and no other was employed. The excess of blood going to his brain was directed to external organs by massage, and the patient was thus bled into his own peripheral circulation. He continued well till the day of his death from cerebral hemorrhage, seven months afterwards.

After being repeatedly desired by two old hemiplegics to give them more than thirty to forty minutes of massage, I finally yielded, with the result of fatiguing them unduly for several days. In two cases of chronic myelitis I used massage for several weeks, without either benefit or harm resulting. In a case which was diagnosticated by an ophthalmologist and a neurologist as one of tumor of the brain, the first massage of the head relieved the pain which was at its worst and put the patient to sleep for several hours in the absence of his usual large dose of morphia. The sleep was longer and more refreshing than that which was obtained from morphia, and so deep that the patient did not know when I left the room, though I spoke to him in an ordinary tone. Subsequent massage and everything else had less effect, and the patient died. In another case diagnosticated as chronic meningitis at the base of the brain, the severe pains that radiated from the occiput to the

---

[1] One of those best acquainted with chorea. — Von Ziemssen.

temporal region in the early stage of the illness were repeatedly relieved by massage, so much so that it was thought probable recovery would follow — a hope that was in vain.

In hyperæmia of the brain and its membranes, whether owing to an increased flow of blood to this region, or a hindrance to the return of a normal quantity, stroking the neck so as to hasten the current towards the heart in the jugulars, has a rapidly depletory influence similar to copious blood-letting or compression of the carotids, but without the possibly injurious effects of these. The pressure of blood in the cranium can thus be quickly lowered, and it has proved an excellent preparatory measure for the use of other depletory agents, such as cathartics, etc.

Gerst, military surgeon at Würzburg, has obtained notable advantages from the use of massage in two cases of serious injury of the head. In one of these there was concussion of the brain with contusion of the soft parts of the cranium, hemorrhage externally and in all probability internally. The other suffered from a fissure of the cranium with a contusion of the left half of the chest and effusion of blood into the pleural cavity. The wounds were treated antiseptically. Stroking (*effleurage*) of the neck, so as to rapidly empty the jugular veins, was employed in both cases and "this exercised a salutary action in preventing the evils of hyperæmia, and in favoring the absorption of the exudation. Thanks to this influence, we have seen the symptoms of compression of the brain produced by the intra-cranial hæmatoma, as well as that at the seat of the extra-cranial hemorrhage, ameliorated rapidly. There was, in one of the cases, paresis of the iris of the left eye with deviation of the tongue to the left and paralysis of the bladder and of the rectum. These all disappeared quickly, and there were no symptoms of inflammatory reaction on the part of the brain. It is certain that the effleurage has been useful in these two cases ; for at each application the patients experienced marked relief in the head. Their pains diminished and the paralytic phenomena disappeared." [1] Gerst adds that for a long time he has thought that advantage might be derived from massage

---

[1] Schmidt's Jahrbücher, 1879, No. 10, p. 73.

in paralysis of cerebral origin.   He used it for a few minutes at a
time, and several times daily.   Relief followed from this which
was not apparent from the use of ice-bags on the head.   Brant-
ing and Georgii hoped by this means to maintain the nutrition
of the muscles and to prevent the consecutive atrophies and
contractures.

Prof. Erb tells us[1] that external frictions are much used and
highly esteemed amongst the non-medical public in Germany,
but are usually rejected by physicians; and in this respect, he
says, medical scepticism often goes too far.   Of the efficacy of
external friction in diseases of the spinal cord and its envelopes
he believes he has *quite accidentally* proved to himself the benefits of such procedures, and he is therefore unwilling to see them
quite abandoned.   " Friction with spirituous substances upon
the skin," he says, " may excite and enliven the action of the
spinal cord, and bring to pass a better functional condition and
nutrition in it.   The soothing effect upon the peripheral cutaneous nerves produced by inunction with warm oil or narcotic salves has a soothing action upon the central nervous system, and this contributes to the removal of diseased conditions.
Moreover, they sustain the courage of the patient."   It is evident that Prof. Erb has not yet accidentally proved to himself
the tonic and sedative action of massage without the aid of spirituous liquors, oil, or narcotic salves; and when he does, it may
puzzle him to decide when to use massage and when to use galvanism.

My friend, Dr. David F. Lincoln, wrote to me inquiring: " Do
you consider massage safe in cases of vascular degeneration of
the cerebral arteries ?   Do you see any contra-indication for it
in cases of senile degeneration of the brain accompanied by
slight paralytic shocks (thrombosis)? "   To these questions my
reply was, massage contra-indicated.   Soon after this I met the
doctor, and he told me that the patient to whom his questions
referred had died since he wrote to me.   If massage had been
used in this case it would probably have been credited with
some of these shocks, and perhaps with killing the patient,
according to Dr. Althaus.

---

[1] Ziemssen's Cyc., Vol. xiii., p. 187.

CHAPTER XI.

MASSAGE IN WRITER'S CRAMP AND ALLIED AFFECTIONS,

WITH A REPORT OF TWO HUNDRED AND EIGHTY-FIVE CASES (TWO HUNDRED
AND SEVENTY-SEVEN TREATED BY WOLFF).

. . . "He sweats, strains his young nerves, and puts himself in posture."
—CYMBELINE, iii., 3.

OVER-USE of muscles and nerves, especially in fine work requiring a high degree of delicate co-ordination of individual movements and voluntary impulses, as in writing, sewing, knitting,
watchmaking, playing the piano, harp, or violin, etc., gives rise to
similar disturbances. So do also, but less frequently, excessive
use of muscles in heavier occupations, such as painting, telegraphing, tailoring, shoemaking, blacksmithing, milking, etc., occasion
like troubles of motion and sensation. Predominance of symptoms may be of a spastic, tremulous, or paralytic form, with extreme fatigue, pain, formication, hyperæsthesia or anæsthesia,
and thrills like electricity. There may be total inability to
perform the accustomed movements, or if they be attempted for a
few minutes, the symptoms just named appear. The spasms may
be of flexors or extensors; there may be rigidity or contraction
of the muscles, local or general tremor. No two cases are exactly
alike, as these symptoms are variously combined and usually only
called forth on attempting the work that has brought them on,
while for all other purposes the hands and arms are well. As I
predicted some time ago, we can now add another form of cramp
to the list, namely, manipulator's cramp, as the penalty of those
who do massage without knowing how, and the sufferer supposes
that the trouble in his arms is owing to his having imparted so

much "magnetism" out of them to his patients — his conceit not allowing him to think that he is only suffering from an unnatural, constrained, and awkward manner of working.

In recent and slight cases good results, though few, have been obtained from galvanization, but the prognosis in general, from any treatment whatsoever, has hitherto been regarded as unfavorable, unless some objective points can be discovered as the source of the malady, such as neuritis, painful scar, or bad writing materials. In the New York *Medical Record* for April 28th, 1877, I published the following cases and remarks. The first was taken from *Hygeia*, July, 1873, by *Schmidt's Jahrbücher* for 1875, and the next two from Virchow and Hirsch's *Jahresbericht* for 1874. In April, 1873, a large, strong, healthy man, thirty-two years of age, by occupation a secretary, consulted Professor Rossander for writer's cramp. Two years before he had the first symptoms, which appeared as fatigue after an hour's writing. Later there was absolute impossibility for him to write at all; he could only hold the pen and make a few strokes, and on attempting to do more the hand was drawn up off the paper, and this became more violent and was accompanied with pain. The hand and arm were quite strong and normal in every other respect. The treatment consisted in the use of massage twice daily, energetic kneading of the muscles of the hand — of the thenar and anti-thenar, of the interossei and lumbricales; and with a small wooden cylinder percussion of the muscles of the thumb and little finger, and also of the forearm, especially of the pronators and flexor and extensor carpi ulnaris. At the beginning of the treatment the thenar muscles, on being beaten, contracted, but not strongly; but the abducens minimi digiti did not contract. Later, by degrees, it did. Subcutaneous injections of nitrate strychnia, ten to twelve drops of a one-per-cent solution, were also given daily in the ulnar side of the forearm. After one week, there was marked improvement, and after four weeks of this treatment the patient was well.

Drachman's case was that of a lady, sixty years of age, the Countess D., who for eight years had suffered from writer's cramp, with tonic convulsive spasms of the upper arm, as well as of the forearm. She could neither write nor take hold of small objects with the right hand. On the middle of the flexor side of the forearm, in the tract of the median nerve, there was felt deeply situated a small, spindle-shaped, smooth tumor, pressure on which gave rise to pain, and on strong pressure there was produced severe pain in the fore and middle fingers and thumb. Many different kinds of remedies had been tried in vain, and amongst these electricity, baths, and embrocations. After two months' treatment with massage, without the use of any other means, the patient could write and do all kinds of fine handiwork without fatigue. The neuroma had decreased and could scarcely be felt.

Gottlieb's case was that of a woman, fifty-two years of age, who had suffered from the malady for nine years. She had been obliged to write nine hours daily, for two years, at the end of which time the pen fell suddenly out of her hand. At each attempt afterwards to again hold the pen, the hand trembled strongly. The forefinger of the right hand had become quite incapable of holding the pen, and the middle finger was also similarly affected, but to smaller degree, on account of which the thumb, supported by the fourth and fifth fingers, were brought into use. There was no pain, but a feeling of formication on the dorsal surface of the two affected fingers. Upon attempting to work with the left hand, the same symptoms were brought forth, but in a milder form. There was œdematous swelling of both upper extremities, also in the first and second metacarpal spaces; and extending upwards between the muscular interstices of the whole arm, were found here and there spots of infiltrated connective tissue, which were painful on pressure; these were more marked on the right side. There was considerable anæsthesia of the second and third fingers of the right hand. After thirty-seven massages the patient was discharged; the hands and arms were perfectly normal; only on long continued and forced writing was there felt the least fatigue.[1]

It is not stated in the last two cases in what manner the massage was used, but in all probability kneading or *pétrissage* and percussion with passive motion predominated over simple upward friction.

---

[1] I have since had access to a more full report which states that she could, two months later, write twelve hours a day.

When sufficient time for rest has been allowed, and in the absence of spasm, or spasm of the flexors alone being present, I should think it might be useful to add resistive motion so as to bring systematically into more powerful action the opposing and less-used extensors, which would tend to restore harmony of action by a counter-balancing distribution of will, nerve, and muscular effort.[1]

The indications for the use of massage in such or any other cases could not have been better laid down than has been done by Althaus in the following words : " A really effective treatment of scrivener's palsy must be an agent which is at the same time both tonic and sedative in its neuropathical effects, which must have the power of restoring the circulation of the blood in the suffering parts to its proper condition; which is capable of promoting the absorption of serous effusions, and will thus cause the nutrition of the maimed ganglia to be raised to a normal standard. Such an agent we possess in the constant current "; and I would add, in massage also. When neither the constant current nor massage alone did any good, it might then be well to try both ; on the principle of *shot-gun* therapeutics, perhaps, it would be better to combine the two on starting. " By stirring up the nerves and muscles of a limb, you may, says Reynolds, " to a certain extent, act upon the other ends that are in the brain and spinal cord, and so improve by careful usage the nutrition of the brain and spinal cord." How exactly this agrees with the statement of Professor Erb !

An old gentleman, a lawyer by profession, at times suffered from lumbago, brought on by over-work, for which he sought relief by massage. While *masséeing* him for that more than once, he called my attention to the thumb, fore and middle fingers of the right hand, the fatigue of which from writing he was fond of designating writer's cramp, as it was only after several days' rest that he could resume the use of his pen. On two such occasions, immediately after prolonged writing, I manipulated his

---

[1] I fear that these cases and the suggestions made in this paragraph (in 1877), are rather damaging to the claims of priority made by those who have since published their experience with the use of massage and gymnastics in the cure of writer's cramp and allied affections.

hand and arm thoroughly, and the following day he could use his pen as freely as usual.

A middle-aged gentleman, of vigorous constitution, a lawyer in extensive practice, was frequently obliged to employ an amanuensis for several weeks, in order to recover by rest the ability to use his pen. For the general weariness at such times, when he could, he had recourse to *massage*, and at two different times while treating him thus, I gave the fingers, hand, and arm more thorough and special malaxation, with the result each time, after two sittings, of his being able the following day to resume his writing with ordinary ease.

Had these two cases not had the intelligence and the means to rest on the first note of warning, they would doubtless have become confirmed cases of writer's cramp. There are many cases of this kind in which I have had an opportunity to try *massage*. No objective symptoms were present, and as rest was indicated more than anything else, I did not trouble them with resisting or acto-passive motion.

Mr. B. is a gentleman of leisure, in fine health, and of unusual muscular vigor. His favorite pastime is music, and he is a skilful pianist, fond of playing the most difficult pieces ; but for years the forefinger of the right hand has been a distressing bane to him, for without warning it will, as it were, *miss fire*, flex toward the palm without striking the key note, and then he has to desist for the time, sometimes for several days. In one week I gave him three *massages*, manipulation, percussion, and acto-passive motion, but with no benefit, and I doubt if further treatment of the same kind would have been of any avail, as the finger, hand, and arm were perfect and powerful in every other respect. He had tried electricity, rest, and gymnastics.[1]

Over-use of any group of muscles gives rise to similar disturbances, some of which may be relieved by *massage*, as the following cases will help to show. H. W., æt. 25, enjoys good health, and has strong muscles ; by occupation a pianist and astronomer. For a year past, June 23d, 1874, his wrists have been weak and

---

[1] I saw him several years afterwards, and the same defect still existed, but he had got used to it.

lame, which he attributes in great part to the frequent and forced efforts required in elevating and changing the direction of his large telescope, which strains the extensors of his hands very much. He can play but fifteen or twenty minutes on his piano before his fingers and wrists give out from fatigue and ache. No visible or tangible defect could be found save a somewhat constrained, stiff-bent position of the fingers, making voluntary extension difficult and disagreeable.

The treatment for several months had been rubbing with liniments and half a dozen layers of bandage wound around each wrist, without any improvement resulting. These were left off when *massage* was begun, June 23d, 1874. The first four visits were devoted solely to manipulation of the fingers, hands, and arms. I find my notes quote Mr. W. as saying that his hands and arms felt stronger after the first handling. At the fifth and subsequent *massages* I added percussion and resisting motion to all the natural movements of the fingers, hands, and arms, but more particularly to *extension* of the fingers, and of the hands on the forearms, and this was carefully kept within the limits of the patient's strength, so that at no time should he be made painfully conscious of his disability, as this would have frustrated the object of the treatment. In thirteen days from his first visit to me he had eight massages, at the end of which time I again find my notes quote him as saying that " if any one had told him that his wrists and hands could have been made so much stronger as they now were in so short a time, he would not have believed them." He could then elevate and move his telescope about with ease, and play on his piano for an hour at a time before fatigue came on. Massage was continued for a few weeks longer, and the patient got quite well, so that he could use his upper extremities *ad libitum* for any mortal length of time. He has continued well, and for his scientific attainments he has recently been employed by the United States Government in a situation requiring a man physically perfect.

Dr. F. E. Corey, of Westboro', Mass., very kindly sent me the following interesting account of a case of over-use of the muscles which move the humerus backward and upward, which he treated successfully with massage:

Mr. D. C. B., æt. 66, by occupation a cutter of leather for boots; has worked at this business a long time, following a pattern with his knife by just the same motion day after day. Previous to my acquaintance, he has had attacks of lameness in the right shoulder, which have obliged him to discontinue work for weeks at a time. The lameness for which I treated him commenced last winter by a slight pain on making the cutting motion, and it slowly increased in severity until the movement could no longer be made without a degree of pain, which led to a discontinuance of his work.

March 23d, 1875, he called at my office, and I found that the posterior fibres of the deltoid and the external head of the triceps were considerably rigid and tender on pressure, and the pain of motion was referred mainly to them, though at times the teres muscles seemed to be involved. I began the *massage* that day, working in the direction of the fibres involved. After about an hour's manipulation my patient declared that he could move the arm better than before, and expressed the belief that this was the proper treatment for him. For about a week I applied the treatment every day, always with the assurance of progress from the patient. After this the application was made at longer intervals until the 19th of April, when my patient could discover no traces of his lameness, being able to carry the arm in every direction without pain. There has been no return of the trouble up to this date, August 11th, 1875.

In two weeks from the commencement of treatment he returned to his work before he was entirely well.

But of all the vexatious cases that have come to me for massage in an experience of twenty years of this treatment there have been none more trying than confirmed cases of writers' cramp and allied affections. It was not with regret on my part when they gave up a conflict, the result of which was so doubtful. The history of these refractory cases in my hands has, in general, been improvement after a few massages, soon followed by relapse, abandonment of treatment, and resignation to their fate. Did I graduate the massage and exercises in quality and quantity to meet the indications of different cases? was a question that often occurred to me, and which was as difficult to answer as experts in diseases of the nervous system doubtless find in formulating the best methods of applying electricity in such cases. It was with feelings of satisfaction to me when at last a case came that was by no means indifferent to his fate, whose penmanship was his livelihood and that of his family, and whose faith had not yet been shaken by the croakings of wiseacres, and who had not been deceived by amateur rubbing dignified with the name of massage.

This case was Mr. A. J., thirty-one years of age, who was referred to me by Dr. Geo. W. Gay on the fourteenth of January of this year. He was in good general health, and his muscles were well developed. It was two years before this that he first observed that he was not writing with his usual ease and accuracy, as if out of practice. He is a professor of writing in a commercial college. He gradually grew worse, so that he had to use a larger pen-holder, and grip it harder and harder. Occasionally there were days when he could write well and easy. It was just after doing some very fine writing that had to be reproduced, and which he first outlined in pencil, that his difficulty began. When he first came to me he could write a few lines well and naturally, then the hand and arm became tired, the hand jumped and trembled, he grasped the pen more firmly, and as the fingers contracted he lost his grip altogether; so that he presented three phases of writers' cramp — tremulous, spastic and paralytic — in one or more of which it it usually occurs. When well, he wrote with his hand in the so-called regulation position, resting on the tips of the little and ring fingers, but gradually he had to let his hand descend so as to write while resting it on the whole of the middle phalanx of the little finger, and using the muscles of the forearm rather than those of the hand and fingers. At times the forefinger alone would jump from the pen-holder, and then he would hold it down with the thumb and endeavor to continue writing.

Examination of the hand revealed almost nothing — apparently slight stiffness of motion in the interossei between the metacarpal bones of the index and middle finger, but not more than is often met with in those not troubled with writers' cramp. There was, however, not full strength in extending the fingers, which would point to over-use of the flexors, and the need of exercise of the extensors to counteract this.

It was not till after I had seen our patient a few times that he told me that nine years before he had sprained his back by attempting to shut a heavy trap-door in a steam-boat. He was beneath it, with his hands and arms extended over his head, when the boat gave a lurch, and he was suddenly thrown backwards. For this he had constantly worn a corset which enveloped his whole trunk, in order to support his back. With this he was comfortable, and did not require to lie down to rest during the day, but without it he drooped and sagged down, and soon a burning spot appeared about the middle of the dorsal region. Examination proved that there was nothing at all the matter with his back, unless it were muscular weakness, due to having worn the corset too long. After two massages the patient felt as if he had a new back, and could go for half a day without his support, and in the course of two or three weeks it was laid aside entirely. If the condition of his back had anything to do with his trouble in writing, the latter ought to have appeared much sooner. Neither do I think that imagination had anything to do with his writing, for he did not know what was the matter with him until the day he was sent to me.

To keep the patient at his work, and at the same time attempt to get him well, was the problem to be solved. For home exercises I prescribed at first active extension and separation of the fingers, and later the same against resistance by means of rubber bands and tubes, so many movements at stated times, in order to bring into greater action the less used extensors, and also to give a change of exercise to the interossei, and thus help to restore the lost equilibrium of will, nerve and muscle. But to prescribe writing exercises for a patient whose chirography was like copper-plate did not seem so easy a matter. However, I had no difficulty, for it was evident that he was painfully slow and particular, and when fatigue came on after a few lines he had hitches in rounding the backs or left lower curves of his l's and e's, and in making the upward stroke of the leg of his g's. Therefore, for home exercises in writing I directed large l's made quickly and continuously, followed by the reverse of these, making m's, so as to make him write from the upper arm and shoulder. As time went on we gradually reduced these in size, so as to bring more into play the

muscles of the forearm and hand.   When he had become proficient
in these, the next exercise was a little more difficult, and consisted
of *lelelele*, large and rapid at first, then gradually diminishing, and
later the exercise was *legleg*, practised in the same manner many
lines at a time, and in this way he soon got over his hitches and
halts.

But calisthenics and elementary writing exercises, though help-
ful, have never been known to cure a case of writer's cramp without
other assistance.   And for this purpose I gave the fingers, hand,
and arm massage, deep manipulation, almost daily for four weeks.
After the first two massages the patient wrote with unusual facility,
but tired as soon as usual.   After the third massage he was fatigued
at the end of the first line, and it is a wonder he did not give up
treatment then, as these cases are apt to do.   After four massages
he wrote with greater ease, and made delicate movements of fingers
and thumb which he had not been in the habit of doing, and he
was but slightly fatigued with ten lines.   After the third massage,
which included the back, he was almost faint with hunger, though
he had just had dinner before coming to me.   I have observed the
same effect in other cases, in one a physician, from percussion alone
for a few minutes on the back.   At the fifth visit there was some
lameness of the muscles of hands and arms from the manipulation,
which had not been rough, and this is generally a good omen.   He
thought the writing exercises which I prescribed for him were
excellent practice to train his boys at the commercial college to
write a free, easy, and rapid hand, so he used them for that purpose.
After the fifth and six massages, wrote still more easily and for an
hour and a half each time, stopping occasionally to explain to his
students.   At the end of nineteen days he had no difficulty in
grasping his pen-holder, and he could write with ease for three
hours, and at the end of twenty-eight days he wrote with ease and
fluency and animation.   And thus he improved, with variations,
but all the time making a better average.

At times we had to call a complete halt for a few days in his
home exercises, when it was evident that he was overdoing and
getting his nerves and muscles into an irritable condition, which
was relieved by massage alone.   But when this condition has

arisen of its own accord or from writing, in other cases, it might be an indication to urge them on with exercises in order to tire out the affected nerves and muscles and their central connections, and thus allay over-excitability. The same means incites nerves and muscles that are inactive, but here, in order to be of benefit, must stop short of over-exciting them.

Our patient might have been discharged at the end of four weeks, but this was not in accordance with his wishes, for he did not then feel safe without the aid of massage, so he continued to visit me two or three times weekly for several weeks longer. At the end of six weeks, though he was generally fatigued from sickness and death in his family, yet he had not the slightest difficulty in giving his writing classes full instruction from nine to twelve o'clock, and it was during the last ten days of this time that I thought it well for him to have a tonic consisting of five minims of tincture of nux vomica, twenty minims of cascara cordial, with thirty-five drops of elixir of calisaya, three times daily. He called upon me again ten weeks from the time I first saw him to report that he had attained perpetual motion, for the longer he wrote and the more he exercised, the easier it became and the better he felt. I have heard from him recently, and he has continued well. Without this patient's hearty coöperation, he would doubtless have sunk into the slough of despond.

The two following cases are the kind not likely to be benefited by massage nor anything else : —

Mr. W. was forty years of age, well nourished, and had good, strong muscles. When eight years of age he had scarlet fever which left him with general neuralgia, from which he has never fully recovered. He was a clear-headed man of business, though he suffered from dull headache all the time, slept poorly, and woke up tired. He was not so well when on a vacation for a few days as at business. He had general feelings of burning, fatigue, and stiffness, and also heaviness of the legs. There was literally too much tension of both mind and body. He was evidently a pronounced case of neurasthenia. From the time he learned to write until he was eighteen years of age he wrote a large free hand, but at this time he became a clerk, and wrote a small, careful hand, slowly, and with increasing difficulty. In January, 1883, he was much run down, and work was hard for him, and during the following month he had to stay at home with indefinite symptoms of prostration and fever. When he returned to business he could not write at all in the usual way, but took the pen between the index and middle fingers. At first there was flexor spasm of the fingers and thumb, accompanied with extensor spasm, in dorsal flexion of the hand. After a time he was obliged to give up writing nearly altogether, so that by September of the same year he had to limit himself to the signing of his name. More than this caused the hand to tremble, the fingers and thumb to flex, then the hand would curl up in the wrist, and the cramp became so painful that he had to desist. Counting coin produced the same symptoms in the left hand. He suffered most from discomfort and tension throughout the whole of the right side of his body; his right eye was incorrigibly astigmatic, and he had catarrh of the middle ear as long as he could remember.

He went to Europe and visited an eminent specialist twice a day for a month for massage, writing exercises, and calisthenics. He came home labelled " cured," but he could not write any better. As he had previously been to me for his general troubles, my method of doing massage stood still higher in his estimation for his visit to Europe. So he came to me for a special trial at his writer's cramp. Under a month of daily massage with the use of Jacoby's wristlet and elastic tubes to exercise the extensors of his thumb and fingers at home, and practice with Nussbaum's writing apparatus, no benefit ensued. His business was in a bad state, and he was under constant worry. This is one of the cases not likely to be improved by massage, or if so, only temporarily.

Mrs. F. H. was forty-nine years of age, and for the previous twenty years had written much. Three years before coming to me she first noticed that it troubled her to hold her pen. There was nothing peculiar about her handwriting, but after a few lines the thumb slipped off the pen-holder by gradually extending itself (clonic spasm), and this was attended with pain in the metacarpal space between the thumb and index finger, and at the outer aspect of the insertion of the biceps, in the region of the musculo-spinal nerve, as well as in the anterior fibres of the trapezius and at the base of the skull. With the pen between the fore and middle fingers she could write fairly well for ten minutes, but more than this caused pain in the places just mentioned. The whole arm felt lame and heavy. Housework, such as dusting or ironing, was difficult, and every motion of the arm caused pain. The left arm had been similarly affected for three months. After I had seen her a few times, she admitted that she felt somewhat ashamed of the unusual but natural prominence of the metacarpo-phalangeal joints of the thumbs, and in order to partly conceal this, she had been in the habit for many years of moving this part of the right thumb into the palm of the hand. Hence, no doubt, had arisen spasm of the adductor pollicis, and on examina-tion it was found that this muscle was sore and tender. Her gen-eral health had always been delicate. She slept poorly, was dys-peptic, constipated, had uterine catarrh and various other ail-ments. She had been utterly proof against all sorts of medication in the hands of the most skillful gynecologists and neurologists. Massage had no better result, but with a Nussbaum's apparatus she gets along very well.

I might weary you with such doleful accounts of unsuccessful cases.

But with quite recent or incipient cases of writer's cramp and similar troubles, my experience has been much more cheerful, the patients recovering safely, quickly, and pleasantly with a very few massages, so that they remained in blissful ignorance of how near the rocks and shoals of disability they had been sailing. I will mention one. It was that of Miss E. P., twenty-eight years of age, of slight frame but tolerably firm muscles. The previous winter her arms and hands would frequently get tired from play-ing the piano, but would recover by resting for a day. She had the same experience the following winter before she came to me. Three days before her appearance she had played the piano for three and one-half hours, too long for her, with the result that the hands and arms were greatly fatigued, and there was twitching of the muscles, and rapid slight contractions of the fingers, alternating with a dull ache, and the whole arms felt lame to the shoulders. Under two days of rest alone her symptoms had become worse, and on the morning of the third day, when she first came to me, she could not turn a newspaper nor play a single note on the piano, and there was slight swelling of the affected members. Thirty minutes of stroking or *effleurage*, alternating with deep massage, at 10.45 A. M., was accompanied and followed by perfect comfort for six hours, and the slight discomfort which then returned was forgotten in the social enjoyment of the evening. Next morning slight return of symptoms below the elbows, none above. Massage at 3 P. M., with great effect. She returned again after two days' interval and reported that she had played the piano for fifteen minutes the previous evening, and thirty minutes next morning, with but little uneasiness. Massage of one arm, she said, relieved the other before it had been *masséed*. After the third massage she regarded herself as cured, and said that she would certainly return if she were not. As I have not heard from her, she probably fully recovered.

In this last case the advantages to be gained by months of rest were evidently made much more secure by three massages in a few days.

All the heavy armor of medical and surgical therapeutics have accomplished little or nothing in the way of slaying this Philistine which defies us in the form of writer's cramp and cognate disturbances. It required a David to meet this giant, and he has come forth as Herr J. Wolff, who, declining the lightning, the sword, and the resources of alchemy, has chosen but two long despised stones from the river of therapeutics, namely massage and gymnastics, and with these in his sling he wields it with such skill, vigor, and enthusiasm that he slays the giant at almost every blow. True, David used but one stone, and the analogy does not suffer when we learn that gymnastics alone will not suffice. The use of massage and movements in writer's cramp and similar defects was well known before the time of Herr Wolff, but their superior value and efficacy were not fully demonstrated until he became interested in them and until his method and results were published by Dr. R. Vigouroux in *Le Progrès Medical* of January

21st, 1882, and by Dr. Th. Stein in the *Berlin. Klin. Wochen.* for Aug. 21, 1882, about five years after my paper on the subject in the New York *Medical Record,* which had the honor of being quoted by the *Journal de Médecine* for October, 1887, and then forgotten.[1]

Herr Wolff used to be a famous teacher of writing, and in this capacity he had pupils sent to him from all parts of Europe to correct their bad chirography. In this manner he became interested in writer's cramp, and he soon improved his naturally keen observation by a study of anatomy and physiology and of disturbances of the central and peripheral nervous system. " The means employed," he says, " are massage and gymnastics combined in a novel way, their effects being mutually antagonistic, and adapted to each individual case. Let us assume, for instance, a pronounced spasm of flexors. Here the muscular antagonism is physiologically altered, there is spastic contraction of the flexors and abductors. In proportion as this continues the sense of weakness increases, and the hand-writing becomes more uncertain. As the contracture ceases the weakness disappears and the cramp is cured. In order to definitely find the fingers affected I perform massage of them,[2] partly centrifugally,[3] partly centripetally, as far as the wrist. I then cause the patient to execute different free motions such as bending and stretching, spreading and contracting, continued for hours until the hand is fatigued, and these are repeated until the patient is able to move each finger voluntarily in all directions. These manipulations carefully repeated have done me excellent service, and the affection in most cases was cured in from three to four weeks."

Dr. Th. Stein says of the method that " it rests exclusively upon active and passive gymnastics of the fore and upper arm, upon massage, percussion and friction of the same parts, and after a

---

[1] When any one not familiar with the past history of massage realizes its benefits in intractable cases, it is to him a veritable discovery and it might be better to allow him to think that he is the first discoverer, for the resulting enthusiasm goes far towards influencing favorable results. Pursue any treatment with lukewarm indifference, and little or no benefit ensues.

[2] The spasm alone would indicate the fingers affected, when massage might not.

[3] Why centrifugally ? It is not advantageous here.

time elementary exercises in writing prescribed and adapted to each case by holding the pen in a definite manner. These are gone through with two or three times daily for half an hour or so at a time. The peculiarity of the method is that Wolff, in consequence of much practice, is able by means of his hands and by elastic bands to fix exactly those muscles which require special exercises, and which we physicians and electrotherapeutists are unable to point out so exactly even with the most minute electrodiagnostic examination."

From 1877 to 1882 Wolff has treated by massage and gymnastics in all 277 cases of writer's cramp and such troubles. Two hundred and forty-five were writer's cramp and 132 of these were radically cured, 22 improved, and 91 without result. Thirty-two were pianist's, violinist's, telegrapher's and painter's cramp; and of these 25 were cured. In all 157 were cured, 22 improved, and 98 not cured.[1] Of the 132 cases of writer's cramp cured, 108 were men, 24 women; 88 of the men were married and 20 single. Of the women 17 were married or widows, and 7 were single. Of the other forms of so-called cramp 27 were men, and 21 of these were married; 5 were women, and 2 of these were married. Most of the women with writer's cramp were widows.

Wolff considers that of those who were not cured, it was in part owing to their want of energy and patience in carrying out the treatment, and in part owing to central disturbances. The cases were from Frankfort, Berlin, Vienna, Munich, Dresden, Freiburg, Amsterdam, and Paris. The results are vouched for by the highest authorities, Billroth, Benedickt and Bamberger of Vienna, Charcot and Vigouroux of Paris, Esmarch of Kiel, Wagner of Leipzig, Bardeleben of Berlin, Hertz of Amsterdam, and Nussbaum of Munich.

Two cases that were sent by Charcot to Vigouroux for electrical treatment were consigned by the latter to Wolff. Both were cured in fifteen days. The first was M. D., æt. 25, robust and vigorous, who had written rapidly twelve hours daily during the

---

[1] (It is not unlikely that of these 98 the progress of the affection in some may have been arrested, of others retarded, while possibly in a few there may have been an apparent aggravation.)

preceding winter. He had been unable to write for five months. There was functional spasm of the long flexor of the thumb, first interosseous and external radials; increased excitability to electrical and mechanical impressions, and also of tendon reflex. The patient was anæmic and slept poorly, but these were overcome by static electricity without improvement of the hand. Galvanization, and later galvanization and faradization, were used for three months without relief. Cured in fifteen days by Wolff. The second case was M. F., draughtsman, 27 years of age and robust. He used the left hand easily for writing and drawing, as he had suffered three years from spasmodic movements in the right hand if he attempted writing, drawing, turning the leaves of a book, or twisting his moustache. At the end of fifteen days' treatment by Wolff he wrote and drew rapidly and easily in the presence of Charcot and Vigouroux. Wolff freely shows his method to physicians, but he cannot transmit his experience, skill, and medical instinct.

Dr. Th. Stein says he has been convinced of the amazingly valuable results of this method of treating writer's cramp by the observation of ten cases which he sent to Wolff. The outline of five of them cured within four weeks is given, and the results in the others were similar. A. R., aged 45, strong and muscular, had suffered with writer's cramp for ten years. It began mildly with slight pain in the forearm and finger-joints, but he continued to write for two years, at the end of which the pain was concentrated in the thumb, fore and middle fingers, and became cramp-like on attempting to write, and finally he could not hold the pen at all. Even trying to sign his name occasioned great straining and nervous excitement, so that perspiration broke out. Divers treatment had no effect. He was assigned to Wolff, who treated him twenty-six days, and then he could write swiftly and beautifully. Six months later he was still well.

Case II.—J. B., æt. 42, tall, slim, pale, but well. Right arm very thin, pain in the upper arm to the shoulder; complains of tension and fatigue through the whole arm. He can do everything but write. On attempting this the hand turns from left to right at the wrist, and the upper arm presses itself involuntarily against the side of the chest. The patient writes one line clearly, the second less so, and at the end of the third the cramp-like feeling comes on, and the upper arm resists powerful attempts to pull it from the side. Any further trial of writing causes him to tremble all over and seem greatly excited. After a rest of five minutes, the same symptoms came on at the end of the first line with acute pain in the arm and hand. Dr. Stein turned him over to Wolff, and he was astonished at the end of three weeks to see the patient looking better, fresher and more cheerful, and perfectly cured. He wrote with ease and speed, and the arm was firmer and stronger.

Case III.—M. S., æt. 27, thin, tall, and feeble, has suffered for four years from writer's cramp. Writing had always been difficult and tiresome to him. After great physical exertion four years previously, he could not write so well, and there was increasing weakness in both arms. By the advice of a physician, he went to a gymnasium and the exercise increased the strength of his arms and hands, but the writing still became worse. An attempt to write with the left hand brought on the same trouble as in the right, which was upward spasm of the hand. Pressing it down brought on trembling, and changing the position of the pen was of no use. After four weeks of Wolff's treatment he was well, writing better than ever, and in his former position, that of a clerk.

Case IV.— H. P., 50 years of age, muscles moderately developed and well in every respect, save that he has suffered from writer's cramp for fifteen years. The flexor spasm was so strong that he involuntarily crushed objects. Writing was troublesome, trembling and bad, and of very short duration. He had been treated in various ways for five years without benefit. After five days under Wolff, there was gradual increase of flexibility of the arm and less contraction of the fingers. Cured in twenty-four days, and well a year later.

Case V.— M. L., æt. 26 years, had suffered for a year from writer's and pianist's cramp. Patient of medium size, well and moderately strong, no nervous troubles in her family. She complains of pain in the upper arm, wrist, and fingers, felt particularly on awaking. Upon playing the piano for five minutes both arms are so tired and languid that she has to stop. She cannot write at all. During the first eight *séances* only gymnastics were used, but with no benefit. After this Wolff *masséed* both arms three times daily, half an hour at a time. The pain gradually disappeared. Then massage and gymnastics were combined and occupied most of the day, and after fourteen days more she was cured of her malady and could write and play on the piano continuously.

Dr. Stein concludes by saying that only those physicians who devote themselves to massage and gymnastics could succeed in obtaining such excellent results, as they have done in other affections of joints and muscles. Neither Herr Wolff nor the doctors Schott, says he, can be vindicated in their claims of priority for the first idea of the treatment of writer's cramp and such defects by massage and gymnastics, however perfect they may have become in carrying these out. Weiss, Podrazky, Zabludowsky, and Cederschjöld also report favorable results from massage in writer's cramp, though their experience has been but limited.

The result of this treatment by Wolff in cases other than those in which spasm and tremor are most prominent, only appears in the case No. V. of Stein, which was unusually obstinate. In three of my own cases, the results of massage and movements, and subsequently the use of electricity and injections of strychnia were unsatisfactory, though the affected limbs became stronger and uncomfortable feelings disappeared. They all arose from over-use. One was inability to write, another could not play on the piano, and the third experienced great difficulty and discomfort in telegraphing. They presented the usual symptoms of such affections, *except the spasm and tremor.* So far as we can learn, these two symptoms were the most marked in Wolff's and other successful cases; it would be of interest to learn if they were absent in his unsuccessful ones, as I suspect may have been the case.

The advantages of massage and gymnastics in the majority of cases of writer's cramp and allied affections would seem to be removal of painful fatigue, spasm, tremor, weakness, inco-ordination of motion, feelings of constriction or tension, and disturbances of sensation ; in one case dispersion of a neuroma,[1] in another absorp-

---

[1] So good an observer as Billroth thinks he has seen tumors dissipated by massage.

tion of œdema and infiltration of the connective tissue. Hence, so far as we can judge, this method is capable, in many cases, of fulfilling therapeutical indications of the utmost importance, such as removal of increase and decrease of resistance in the paths of conduction, excitation, and motion; restoration of harmonious co-operation of individual movements of natural conductivity and excitability, as well as of muscular sense and muscular effort; in a word, correction of underaction and overaction of muscles, nerves, and their central reflex apparatus. Impalpable trophic disturbances of the co-ordinating machinery in the central nervous system are regarded as the origin and predisposing cause of writer's cramp and such maladies. If massage excels galvanism in correcting these disturbances, as would seem to be the case, it must indeed be a remedy of rare value and worthy of being used by the most skilful physicians. But wherein lies the unity of Wolff's brilliant results? Evidently it is in tiring out the affected muscles, nerves, and their central connections, thus allaying over-excitability, which manifests itself mainly as spasm. The same means incites nerves and muscles that are inactive, but to be beneficial, evidently must stop short of overexciting or tiring them out. Hence, the necessity of careful diagnosis and tolerable precision in using massage and movements in these cases. When peripheral nerves and muscles are less excitable than natural, there is absent or defective contraction.

It is true of massage as of everything else, that one person may succeed with it when another has failed. Of the galvanic current, Prof. Erb says, "I have thoroughly satisfied myself of its efficacy in writer's cramp and allied affections, though I have not been able to establish the superiority of any one method of applying it over the others. When good effects have followed, the same results were obtained from all modes of applying it to the arm and neck." He might have added that it could be applied in such a way as to be useless or harmful, and so also might massage.

In other irregular actions of nerves and muscles, massage has been found advantageous after the failure of other means. Dr. Beyer, in the Phila. *Med. News* (April 11th, 1885), has reported a case of tonic spasm of the spinal accessory nerve of central origin, manifested by obstinate contracture of the trapezius and sterno-mastoid muscles that had existed for eighteen months, and defied all other treatment, such as nerve tonics and sedatives, electricity, and injections of atropia. The case was effectually cured in nine months by massage and movements.

CHAPTER XII.

"The same nerves are fashioned to sustain .
The greatest pleasure and the greatest pain."

IN neuralgia of milder form, and in what seemed to be the incipient stages of more severe attacks, as well as in old cases of
neuralgia where everything under the sun had been exhausted
but massage, this has been tried, and it has not been found
wanting in favorable results. Used between the paroxysms of
severe neuralgic pains, massage generally lengthens the intervals between the attacks, and lessens the severity of these when
they come on. In all of the cases which I shall refer to where
massage was employed, other means, constitutional and local,
had not been neglected, and it was usually after the apparent
failure of these that this measure was brought into use.

Pain arising from disturbance in the central nervous system
is, as we have seen, frequently relieved by massage, whether
this has any effect upon the cause of it or not. How much more
effectual then ought manipulation to be in peripheral neuralgia,
where the affected nerves can be reached? If the view of
Anstie be accepted, which would explain every neuralgia arising with or without apparent cause, that it consists in atrophy
of the posterior roots of the spinal nerves in which the pain is
felt, and of the neighboring central fibres and ganglionic cells,
we must conclude that the sedative effect of massage reaches
far beyond the region of application. The opinion of Benedict
that at least all peripheric neuralgiæ are due to slight neuritis
does not necessarily conflict with that of Anstie. Either or
both conditions may be present, but slight neuritis would be

222

the more encouraging for the employment of massage. Evidently the less neuralgia is dependent upon disorder in other organs, the better is the prospect for relief from treatment by massage. Disturbances of sensation from too great tension or relaxation of the tissues offer favorable conditions for treatment by manipulation. When neuralgia is not in nerves too deeply situated, and has lasted but a short time, massage is considered by Johnson and others the best remedy of all. In well-marked degeneration of nerves, and when neuralgia is dependent upon mechanical pressure that cannot be removed, we would not expect any result. In the early and late stage of neuritis, massage is indicated; in the early, it would act as a prophylactic, relieving congestion by causing a free circulation in the surrounding tissues, and by pushing the blood out of the distended vessels; in the late, by causing absorption of inflammatory products. When the inflammatory process is on the verge of softening and suppuration, manipulation would be questionable. Impalpable disturbances of nutrition and undue molecular activity or passivity, whether as cause or consequence of neuralgia, are undoubtedly favorably influenced by massage. The repeated mechanical effect of manipulation and percussion upon old neuralgiæ benumbs and lessens the sensibility of the nerve filaments, and gradually decreases it. In other words, nerves that are already in a state of painful excitement may have this reduced by over-inciting them, andt hus wearing out their incitability. A temporary aggravation of pain is likely to occur at first, especially if light and rapid percussion be used; whereas if slow and heavy blows be given an obtunding effect will probably set in at once. Massage, if used at all in recent and severe neuralgia, should be by gentle stroking, firm pressure, and slow, deep kneading which will favorably modify the excitations of the painful nerves. Recent cases would seem to require more time for treatment by massage than old cases. Those in which I have met with failure from massage have been cases where the constitution had been run down and the mind harassed by cares, and where massage had been urgently requested by the patient in the lull between the paroxysms. Under such circumstances

manipulation is likely to reawaken the pain. Neuralgia of anæmia and malnutrition may be successfully treated by tonics, nourishment, rest, and massage.

It has long been recognized, by Anstie and others, as a diagnostic feature of neuralgia that, notwithstanding the tender
spots at various points of the affected nerves where they emerge
from deeper to more superficial structures, firm pressure may be
made on the painful nerves without aggravating the pain, and
very often with the effect of relieving it when rest alone does
not. The wonder is that this hint has not been more utilized
in practice, and that there are not more claims of priority for
the use of massage in its various forms for the treatment of
neuralgia. Dr. Balfour, of Edinburgh, in 1819, implied that he
was the originator of the treatment of neuralgic and other
pains by compression and percussion. The value of these, when
skilfully used, certainly was a discovery, as was the value of
massage in depriving rest of its evils and stimulating nutrition,
and also its benefits in the cure of writer's cramp, sprains, etc.
Such learned authors as Trousseau and Pidoux have unconsciously been led into error by stating in their treatise on
"Materia Medica and Therapeutics" that one Sarlandière was
the originator of massage by percussion. Twenty-five years
before this, in 1818, Percy and Laurent devoted an article to
percussion in their "Dictionary of the Medical Sciences,"
proving its use among the ancients. But they all seem to have
overlooked the fact that the history of percussion extends far
back, even in mythology, for Jupiter once needed his thunder-
bolts which were forged by Vulcan, because the Titans tried,
by piling up mountains upon one another, to scale heaven and
throw him down. Jupiter, finding himself hard pressed, was
seized with a dreadful pain in his head, which caused him to bid
Vulcan to strike it with his hammer. This done, out darted
heavenly wisdom, his beautiful daughter Minerva, fully armed
with piercing, shining eyes, and by her counsels cast down the
Titans. There has been wisdom in percussion ever since, when
properly used.

Nerve excitation and vibration for the relief of pain and other morbid symptoms by means of percussion have recently become of renewed interest and importance from the scientific and successful experiments of Dr. J. Mortimer Granville, of London. Acute and sharp pain he likens to a high note in music, produced by rapid vibrations; while a dull, heavy, or aching pain is similar to a low note or tone caused by slow vibrations. A slow rate of mechanical vibration upon the nerve interrupts · the rapid vibrations of acute pain, while quick vibration arrests the slower ones of dull pain. By thus introducing discord into the rhythm of morbid vibrations, relief or cure in neuralgia is effected. This is accomplished by means of an instrument called a *percuteur*, which is so constructed as to give at will a slow or rapid rate of blows per minute. In a large number of cases, by this means the cerebro-spinal and sympathetic ganglia can be brought under control, torpid centres aroused to action, reflex irritability of subordinate centres subdued, and these centres placed under the control of the higher ones; the vibrations can be propagated along the trunks and into the branches of the principal nerves from their centres of origin, or called into action reflexly by the afferent nerves connected with those centres. In no instance has there been failure to produce activity of the bowels, even when they have been previously obstinately constipated. "In the treatment of pain by vibration, it is necessary to determine, as precisely as possible, the particular nerve-branch in which it is located, and to act upon that branch alone. The treatment will fail if healthy or normally vibrating nerves are mechanically vibrated with those which are in a morbid state. The rationale of the process of relief is to overpower the tumultuous vibrations of the nerve elements within the sheath." Dr. Granville considers the rate of speed of great importance, but admits that it can only be approximated in any given case by trial, for he says that when the pain of neuralgia is not quickly relieved by vibration, it is wrong to continue working at the same rate of speed, as it is likely to aggravate the suffering. Others do not consider the rate of speed as of much importance.

Amongst my early experience with massage, more than twenty

years ago, there came to me an elderly clergyman using two canes for support. He had suffered for many months from sciatica and not a little from the treatment of the same by blistering, cauterization, etc., which had no doubt been of benefit. The affected leg was weak, cold, and flabby, and there was subacute pain in the course of the sciatic nerve and its branches. Massage was applied every other day, and at the end of a week it was surprising that he no longer used a cane, and in two weeks he considered himself well. At an earlier stage, probably, massage would not have been so rapidly beneficial.

In 1878, I attended Mr. T. D., then sixty years of age. Thirty-five years before, he had been stabbed in the left lumbar region. The wound healed and he had no special trouble from it for thirty years, except that he was painfully conscious of weakness in the left side, arm, and leg. For the last five years he had suffered from attacks of burning, boring pain in the course of the left ilio-hypogastric nerve, extending from the cicatrix of the stab to the emergence of the nerve just above the left pubic bone. The attacks were paroxysmal, and came on suddenly, with no premonition, and went off as they came. They lasted from one hour to two days with slight intermissions, and were so severe as to "double up" the patient, and hypodermic injections of morphia had to be used. The attacks occurred sometimes two or three times a week, at others only once in four or five weeks. During these intervals the patient considered himself well, but in the remissions of pain he was peevish and irritable. During the attacks the bowels were distended with flatus, the passing of which, either up or down, afforded relief. The tissues felt lame and bruised in the painful region. When I was called to the patient, he was in the height of an attack, pulse ninety and feeble, respiration hurried, and he looked pale and exhausted. Massage for half an hour gave him great relief. An hour later, he expelled flatus freely by the rectum, and soon after went to sleep and did not wake up till morning (nine hours after my visit). The massage of the back and abdomen was repeated next day and twice a week afterwards for five weeks. I also gave him a pill of iron, quinine, and nux vomica three times daily, and he gained in flesh and strength. A year later there had been no return of the attacks.

Mr S., 60 years of age, in ordinary good health, came to me with supra-orbital neuralgia, which had troubled him for a year in spite of tonics, sedatives, liniments, and electricity. I gave him nine massages in three weeks, and he was so much improved that the slight pain left soon disappeared without further treatment.

J. R. A., æt. 55, a painter, was kindly referred to me by Dr. Denny, from the out-patient department for nervous diseases at the Boston City Hospital. The case was one of traumatic neuritis and paresis, caused by a fall on the shoulder three months before. Motion of all the joints of the right arm, from shoulder to fingers, was stiff and limited, about one-half normal; wrist and fingers swollen. The posterior interosseus and median nerves seemed to be most affected by the neuritis, and the patient was often kept awake by the pain. He was slowly improving before he came to me. In three weeks I gave him twelve massages, applied bandages tightly, and used passive and active motion freely, and at the end of this time he had so much improved that he could resume his occupation.

Dr. Weir Mitchell, at page 271 of "Injuries of Nerves" reports the following case: "A case of contusion of the ulnar nerve became subject to intense neuralgia, the nerve being hardened and enlarged, owing probably to sub-acute neuritis. It was very tender, and no application of electricity was borne with any patience. The patient was relieved by hypodermic injections, but after using many remedies, and at last the actual cautery over the nerve trunk, without altering its size and tenderness, I patiently tried whether by slow and careful manipulations I could enable it to bear pressure. After a course of gentle friction lasting half an hour, I succeeded in my object, and three sittings enabled me to rub and even knead quite roughly the diseased part. After some thirty sittings he could readily bear the use of ice, which before he could not endure, and the nerve was plainly less in size and of diminished density. This very instructive case has taught me in other instances to help the progress of the nerve towards health by like means whenever there are tenderness and sclerotic conditions. I need not say that some tact is needed in slowly increasing the force of the friction and depth of pressure." In this case massage proved more serviceable than the cautery, ice, morphia, or electricity. The advantage of resuming the use of ice is not apparent.

Berghman reports three cases of hopeless neuralgia cured by massage when everything else had failed.[1] The first was that of a woman who had suffered from severe neuralgia of the ulnar nerve for four and a half years. The pain was so violent, and the inability to do the least thing so great, that for sixteen months she had been in a home for incurables. The ulnar nerve was painful in its whole extent, but worst at the elbow. There was no swelling apparent. Massage was used daily, and in six weeks she was free from pain, and soon after was perfectly able to work. After eleven sittings she was free from pain for a whole day, and after eighteen for two days, and so the patient progressed.

A man, 54 years of age, had suffered from neuralgia of the trigeminus in the left side of the face for five years. The pain was very intense, with exacerbations every five minutes by day, and continual disturbance of his sleep by night. After six days of treatment by massage, he obtained intervals of freedom from pain of three-quarters of an hour; in three days more the intervals of relief were two hours, and after ten days the pain ceased entirely.

A single woman, 30 years of age, had suffered for more than two years from the usual symptoms of coccyodynia; pains over the sacrum and buttocks made very intense by pressure upon the extreme end of the coccyx. Massage was used daily, and after eight days she was well and has continued so ever since.

Westerlund narrates the following case (in Schmidt's *Jahrbücher*, 1877). A woman, 34 years of age, of strong constitution, observed, in January, 1874, a dull sensation in the right arm, which gradually became painful, having its seat mainly in the shoulder-joint and extending from there downwards. After rubbing with various substances and the use of galvanization, there was no improvement. The pain grew worse and was of greater intensity at night, disturbing sleep. Three weeks more of galvanization was tried with no amelioration. The pain became more concentrated in the course of the musculo-cutaneous nerve and over the supra-scapular region,

---

[1] Virchow and Hirsch's Jahresbericht, 1874. From Scandinavian medical journals.

and was aggravated by motion or deep pressure. After eight massages of kneading and percussion, there was so much improvement that the patient withdrew from treatment, but relapsed at the end of six weeks. After fifteen more sittings the pain was all gone, and fifteen months later the patient was still well.

Dr. L. Faye reports the following two cases in *Norsk. Mag.*, 3 R. V. 12, 1875. A woman, 25 years of age, who had been pregnant for two or three months, was suddenly attacked with sciatica, and could walk only with great difficulty, and lie upon the affected side. Various remedies had been used without result. After daily massage for a few days the patient could walk without much difficulty, and after eighteen sittings the cure was complete and lasting. It was only after the patient recovered that Faye learned that she was pregnant. If he had known this before, he would not have used massage. In view of this result it is well that he did not know of her condition.

A man had a peculiar sensation in the neighborhood of the perineum, which was sensitive to touch. The nature of the affection was obscure. Hypertrophy of the prostate could not be made out. Later there came pain in the back and sides. The most varied treatment had been of no use. After thirty-six massages the patient was cured. A slight return of unpleasant sensations was removed by percussion.

Dr. Carl Gussenbauer, Professor of Surgery at Prague, has treated three cases of neuralgia successfully by means of massage.[1] One of these was a man, thirty-four years of age, who had contracted the malady by standing on a cold stone sidewalk. He had been treated by several physicians without benefit, only the use of the faradic current having afforded temporary relief. Later Töplitz thermal springs were tried, but did no good. The patient walked as if he had inflammation of the hip joint, but examination showed this not to be affected. In the course of the exit of the right sciatic nerve were found two knotty indurations which were painful to a slight touch. The whole sciatic nerve was also very sensitive to pressure. Electricity was again tried, but the patient declared that he grew

---

[1] Erfahrungen über Massage.

worse. Fly blisters afforded some relief. Massage was applied over all the affected parts of the sciatic nerve, but especially upon the indurations. These disappeared gradually, and with them the pains, and at the expiration of four weeks the patient left the clinic cured. The next case of Professor Gussenbauer was one of neuralgia of the *nervus cutaneus femoris medius* of six weeks' duration, caused by taking cold. In the course of eight days, with two massages daily, spontaneous pains and sensitiveness to pressure had entirely disappeared, and motion had become perfect. In this case there was a painful point where the nerve came through the fascia lata. The other case was also one of sciatica of several years' duration, treated by massage with similar results. The professor says he could narrate many other cases showing the benefits of massage, and believes that this remedy will find its way more and more into medical practice if physicians will only once prove its results.[1]

Norström reports a number of cases of neuralgia treated by massage. Mme. M., thirty-six years old, came to him in January, 1882, for a violent pain of three months' duration, seated in the frontal region and all over the nose, affecting the nasal, frontal, and lachrymal branches of the ophthalmic, so severe as to deprive the patient of sleep. The secretion of the pituitary membrane was notably increased, and a paroxysm of pain occurred every time the patient attempted to blow her nose. Electricity had been used three months without result. Friction with various preparations had no effect. Cured by eighteen massages, and no relapse. Another was a case of cervico-brachial neuralgia extending to the ulnar nerve, of several years' duration, cured by thirty massages. One was a case of rebellious sciatica with myositis of the sacro-lumbar and posterior femoral muscles with contraction of six years' duration, cured in four months. A case of double sciatica of rheumatic origin in a fat patient was much improved by massage at the

---

[1] " Ich könnte Ihnen, meine Herren Collegen, noch manche andere Beobachtungen anführen, ich glaube indessen, dass die bereits erwähnten wohl geeignet sind, Ihr Interesse für die Massage zu erwecken, die nach meiner Ueberzeugung immer mehr Eingang in die ärztliche Praxis finden wird, wenn sie einmal von der Aerzten selbst mit Erfolg geprüft sein wird."

time of publication. Still another was a case of acute sciatica during acute rheumatism, in which the sciatica got well under massage before the rheumatism did under constitutional treatment. A case of sciatica of seven months' duration without known cause, in which blisters afforded temporary relief, got almost entirely well in twenty-three days under massage, and had no relapse. Mme. D. had suffered for a number of weeks from neuralgia of the femoro-cutaneous nerves which was aggravated by change of position. She was cured in ten days by massage.

Dr. P. Winge has reported a case of double sciatica of nine years' duration, which came after the use of firm compresses applied to the legs during bleeding from the uterus, that was cured by massage in three months.

To 1878, Johnson had treated by massage seventeen cases of neuralgia.[1] Fourteen of these were sciatica, and eight were cured, four improved, and two unchanged. Three were neuralgia in other nerves and they recovered. The number of sittings required were from twenty-seven to fifty-two. These cases were recent, the neuralgia not being of long duration.

Zabludowsky reports four cases of sciatica and one of intercostal neuralgia cured by massage (*Deut. Med. Zeitung*, 1884). Numerous other interesting accounts of cases of neuralgia of long standing treated by massage are given, but space forbids their repetition here. Wagner[2] considers massage particularly useful in peripheral paralyses and neuralgias. The affected nerves should be stretched and kneaded so as to aid in the removal of hyperæmia and exudation. In three cases of sciatica he obtained cure in one at ten sittings, in another amelioration after twelve massages, and in a third case of six weeks' duration no result after fifteen *séances*. It is noteworthy that this last case was subsequently defiant to every other means, including the continued current, and lasted four months in spite of everything.

A case of supra-orbital neuralgia was much improved by eight massages.

---

[1] Virchow and Hirsch's Jahresbericht, 1878.
[2] Berlin Klin. Woch., Nov. 6th and 13th, 1876.

In nervous headache and hemicrania massage does equally as well as in neuralgia elsewhere. Schreiber, Wretland, Westerlund, and others also give favorable accounts of the use of massage in neuralgia. But enough has been said.

In *peripheral paralysis* massage proves useful. In one case of facial paralysis of several months' duration, arising from exposure to cold, that came under my care, the patient rapidly improved and soon got well. Gottlieb reports the case of Mlle. N., twenty-two years of age, who came to him the 28th of October, 1873. Three months and a half before, without known cause, there appeared tumefaction and deviation of the left half of the face, which still continued. No cerebral affection nor syphilis was present. No treatment had been used, and she had improved but little. The ala of the nose and the tissues over the superciliary ridge were depressed, but there was no ptosis. There was slight sensibility around the orbit and nostril. There was no spontaneous pain, but sensation of tension in the cheek. The movements of the face were characteristic of facial paralysis. The brow did not wrinkle on the right side; the eye could be only half shut, and the patient could not whistle. There was neither epiphora nor strabismus. The tongue moved naturally. Electric contractility absent. After twenty *séances* of massage the mobility of the muscles of the face had returned to a normal condition and the patient was well, Dec. 19th, 1873.

Berghman and Helleday report a case of paralysis of all the muscles of the forearm, in which, after the first sitting of massage, there was evidence of muscular contraction in the power of slightly moving the fingers. After three weeks of this treatment, motion had perfectly returned. Cause and duration of the trouble not stated.

G. G. was a delicate child, ten years of age. She had been severely beaten by a boy two years before I saw her. Six weeks afterwards, paralysis of the deltoid set in, which still continued. There was no peri-arthritis, contracture, nor muscular stiffness. She had had two attacks of rheumatism, one before and one after the beating. The arm could be elevated by another without resistance. The strength of the hand and arm was good, though the muscles were considerably atrophied.   For three months before I saw her, she had pain in the arm and shoulder, in the back and front of the head, and also in other places.   Massage for a while, in conjunction with cod-liver oil and tonics, did her no good.

Miss P. was 17 years of age when I was called to her.   She had generally been strong, having ridden much on horseback.   Two years before I saw her she had fallen down six steps of stairs, the force of the fall coming upon the right hand, which was forced into extreme dorsal flexion.   Following this there was numbness, swelling of the back of the hand and pain of the whole arm extending over the shoulder.   She had to carry the arm in a sling for two months.   A year later she had recovered so that she could play on the piano for three-quarters of an hour, though the arm did not feel well.   Very soon after this she had typhoid fever, which laid her up for four months.   The hand and arm relapsed, and baffled the attempts of a skillful neurologist and *masseuse* to improve it.   She could play but five minutes on the piano, and then pain and discomfort caused her to stop, when I first saw her.   Trilling caused the most pain and uneasiness (accomplished mainly by the interossei muscles supplied by the ulnar nerve), while heavy notes were easier.   The ulnar nerve above the elbow was tender to pressure, and over the middle third of the radius there was tenderness and induration in the *flexor longus pollicis*.   After a week of daily massage and careful movements she could throw the whole arm upwards and backwards without discomfort in the course of the ulnar nerve, which she could not do at first, and in another week all tenderness had disappeared from it.   The induration and soreness of the long flexor of the thumb did not disappear until the end of the third week.   It was then that I allowed her to resume piano practice, first one minute at a time followed by massage, then two minutes, after which massage, and so on until in three months she could play two hours daily without fatigue or discomfort of any kind.   She has continued well, now 17 months.

CHAPTER XIII.

MASSAGE IN THE TREATMENT OF MUSCULAR RHEUMATISM (MYOSITIS),
MUSCULAR RUPTURE, ELEPHANTIASIS, AND ŒDEMA.

> Our organs become more dry, tough, and rigid and sooner unfit for use by
> prolonged cold methods of hardening which also bring on premature old age
> and speedier dissolution.— Abbreviated from Hufeland's "Art of Prolonging
> Life," 1790.

THE symptoms usually designated by the somewhat vague and
unsatisfactory expression, *muscular rheumatism*, whether occurring
in those who are rheumatic and suffer more or less from rheumatism, or occasioned by injury, sudden or violent strain, excessive
fatigue, or catching cold, almost always disappear in a very satisfactory manner under treatment by massage when the affected
muscles and fasciæ are accessible to touch and pressure. The
term myositis would be more appropriate for the majority of these
cases, and would be in harmony with what we know of somewhat
similar disturbance affecting involuntary muscles, as in myocardi-
tis and chronic metritis; and at the same time it would indicate
more clearly the nature of the malady, and the treatment required.
In recent myositis, massage acts directly in squeezing the congestion out of the affected tissues and promoting absorption of exudation, thus removing hindrance to the circulation, and pressure
from terminal nerve filaments. It also sets free muscular fasciculi
from minute adhesions which are the cause of partial, irregular,
and painful contraction. Indeed all the local requirements are
met by massage, and not infrequently at a single sitting. M.
Martin, a surgeon of Lyons, cured his confrère M. Petit of an
acute lumbago at a single sitting of massage. He says he has col-

lected over a hundred cases of this kind, and recommends that the massage be repeated in order to confirm the cure.[1]

In chronic cases of myositis, where proliferation and induration of the connective tissue have taken place, with secondary atrophy of muscular fibres and consequent interference with motion, circulation, and innervation, massage will naturally take longer time, but the result in promoting absorption and bringing about a natural state of nutrition and elasticity of the affected muscles is no less satisfactory than in recent cases. The feeling of the affected muscles in old cases, where they can be reached by manipulation, is peculiar. It is neither the natural semi-solid condition of relaxation nor the elastic hardness of contraction, but like strands of whip-cord, with here and there hard nodules, and sensitive to pressure. Pressure of the affected parts of the muscles upon adjacent nerves may cause more pain in the distribution of these nerves than is complained of at the seat of the myositis. Indeed, the same pathological conditions may also affect nerves and their sheaths. But alteration of consistency cannot always be detected even in muscles that can be easily felt, and the same subjective symptoms may be complained of in muscles that are weak, lax, and flabby, probably owing to neuralgia of the intra-muscular terminations of sensitive nerves from lack of nutrition. Here it would only be a question of improving nutrition which may be done by massage and exercise. Tension of tissues which are very expansible causes little or no pain. Swelling of the spleen almost always develops without pain, and hyperœmia of the liver usually attains a high degree before the attention of either patient or physician is called to the increase in size. But tension of membranes, ligaments and tissues which are stretched with difficulty excites severe pain.

In 1870, Mr. F. had suffered with severe pain for seven days in the posterior cervical muscles, and in one ankle which had once been fractured. Anodynes had but temporary effect. The first massage was anything but agreeable ; the second afforded sensible relief, and the third on the third day was pleasant and agreeable and followed by complete and lasting relief. The rigidity and

---

[1] Estradère, Du Massage, pp. 108, 142.

soreness of the affected muscles entirely disappeared at the third massage.

Rev. E. B., an elderly gentleman, had been exposed to a draught of air on his back while sleeping in a steamboat berth on the Pacific Ocean. This brought on severe lumbago. When I first saw him, three months afterwards, he complained of a constant dull, tired ache in the lumbar region which made him prefer sitting to standing, and lying down to sitting up. Sleep was disturbed by the backache. His muscles were of the kind that readily respond to massage, neither too hard nor too soft, nor deeply imbedded in adipose. After twenty minutes of deep malaxation with percussion the patient enjoyed a good night's sleep, and next day stood up for three hours and wrote thirteen pages of a sermon. A few more applications of massage were given to confirm and improve on what had been done, and the patient made an excellent recovery.

In the case of a vigorous adult whose muscles were very firm and well covered with fat, who had suffered from myositis of the deep lumbar and abdominal muscles, for several weeks so severe as to confine him to bed, daily massage for four weeks did not have any effect that I could see, other than pleasing the patient very much. This gentleman, several years before, was greatly relieved of a more superficial rheumatism of the muscles and fascia of the back and hips by means of massage.

Mr. J., a large, stout gentleman, has been subject to attacks of muscular rheumatism in the deltoid, trapezius, or lumbar muscles all his life. At one time he was seized severely and suddenly with pain in the lumbar region so that he could not attempt to move. I was called to him the first day of this attack and found all the tissues of his back hard, rigid and board-like, and almost totally insensitive to pressure. Perhaps the lessened sensation to external impression was owing to the already existing severe pain. Important issues depended on his recovery in a short time, and so massage was given twice daily for a week. At the end of this time the muscles were supple, sensitive, and sore to the touch, but the pain had disappeared, and the patient could move about freely. Three more *séances* were given, and the patient has not had an attack since, now over two years, an unusually long interval of freedom for him.

Mr. L., æt. 41, a vigorous lumberman from Minnesota, had suffered from muscular rheumatism for five years. At one time it had been very acute and general, so that, to use his own language, he was "as stiff as a post and as sore as a boil all over." When he came to me he had greatly improved, but seemed to have come to a standstill with a pain and stiffness still left in the lumbar muscles and also in those of the right hip and chest. Sneezing and coughing aggravated his pains. After half an hour of massage the patient declared that the right hip felt freed from restraint, and lengthened two inches, and he could walk with greater ease. He had three more massages with marked improvement in the comfort and elasticity of his muscles. Dr. French, of Dedham, kindly referred this patient to me.

Norström narrates the following interesting case of myositis involving the trapezius and scaleni muscles with indurations, accompanied by habitual headache and reflex migraine cured by massage. Mme. M., æt. 28, had suffered for several years from vague rheumatic pains which ended by concentrating themselves in the head. Galvanization and faradization, quinine, caffeine, and aconitia had been tried, besides hydrotherapy and a season at Aix-les-Bains, with but temporary relief. The pain was rather dull than acute, and situated in the upper regions of the face, and often sufficiently intense to wake the patient early in the morning, and then no better means of relief were found than for her to keep quiet in bed all day. Her head often felt as if it were inclosed by a vice, the right eye red and tears flowing, and at times there were nausea and vomiting and sometimes spasmodic contractions of the muscles of the face resembling tic douloureux. The paroxysms had increased in frequency and intensity. Intellectual efforts or sad impressions were sufficient to provoke the attacks. Her disposition was changed and she sought solitude. She had become anæmic and believed herself incurable, and had scarcely any faith in massage.

Examination showed that the scaleni muscles and the acromial part of the trapezius of the right side were unnaturally hard and

resistant to pressure, and several indurations of the size of a hazel-
nut affected the muscles throughout their depth. The largest was
at the upper part of the trapezius, pressure upon which caused
more acute sensation than anywhere else, and radiated to the right
side of the face, giving rise to pain analogous to that which was
felt at the time of an attack. At first massage was very painful
on these indurations, but at the end of three or four weeks it was
more endurable. As these callosities disappeared under massage
the headache decreased. In five weeks the paroxysms were less
frequent, but it was only at the end of four months, after eighty-
two *séances*, that the patient was well, sleeping tranquilly, looking
cheerful, and in good condition. The muscles had resumed their
former volume and there were only traces left of the largest indurations. Several months afterwards there was no relapse.

Of forty-nine cases in which the situation of indurations was
noted, associated with migraine, Norström found fourteen at the
upper attachments of the muscles of the back of the neck; in
nineteen they were found in the bodies and at the inferior attachments of these muscles; in nine in the muscles of the sides and
front of the neck; in two in the sub-cutaneous tissue of the head;
in three in the temporal region; and two involved the ganglia of
the sympathetic. The majority of these cases were cured, and in
all but one marked improvement resulted under massage.

Dr. Uno Helleday, of Stockholm, has had some very interesting
and satisfactory results from massage in cases of myositis, which
he has given in *Norsk. Med. Arkiv*, VIII., 2, Nr. 8, 1876. The
following is somewhat like the one just related. A patient, 68
years of age, took cold fifteen years before, since which he had
suffered from headache with heat and pressure in the back of the
head. The pain occurred in paroxysms once or oftener daily with
varying intensity, and at times was so violent that he remained
in bed and could not get up. The pain was most intense at the
nape of the neck and radiated from there to the vertex. At times
one side was much worse than the other. The pain seemed to
follow the course of the occipital major nerve. The cause of it
was apparently induration and thickening involving the splenius
capitis, splenius colli, and scaleni muscles. By means of massage

there was at first amelioration and soon perfect freedom from pain. A man, æt. 32, had suffered for four months from pain in the right leg which made walking difficult.   Upon getting up in the morning, or after having been seated for a while, pain and stiffness appeared in the hip, and after a few steps there was such severe pain in the outer part of the leg and ankle that for two months the patient was almost incapacitated from walking.   Dense adipose tissue and firm muscles made an examination difficult.   Nothing abnormal could be found about the leg, but the gluteus medius and the tensor vaginæ femoris were rigid and painful to pressure. Massage of the hip at first aggravated the pains in the leg and ankle.   These gradually disappeared and the patient recovered. Dr. Helleday has recorded six more cases of myositis of equal interest to any given here, in which both the subjective and objective symptoms disappeared under massage.   In one there were small, hard, round places about the size of the end of the little finger throughout both *cucullares*.   The pressure of massage upon these was painful at first and likened by the patient to the feeling of peas being pressed into her back.   In another there was pain in the knee of the right side, but the region of the hip-joint and groin was so very painful and the gait so limping that coxitis was feared, and it was thought at one time that permanent extension would be necessary.   Hip and knee were immovable and the gluteus medius, quadratus femoris, and the sartorius as well as the flexors were swollen and of diminished elasticity, but not infiltrated very hard.   Besides these, many other cases of muscular rheumatism or myositis cured by massage are published by Prof. Gussenbauer, Mezger, Wretlind, Berglind, and others.

Slight rupture of muscular fibres and the resulting myositis can be treated advantageously by massage and bandaging.   Indeed, these measures have now become classical in the treatment of so-called dislocation of muscles, and are recommended in works on surgery by Bryant and others.   But it is very seldom that massage is ever used in these cases, and superficial friction with liniments has to take the name and the place of it.   In 1879, I attended Mr. J. C., forty-five years of age, a large, stout man in good health.   Four weeks before I saw him, he stepped from a horse-car before it had stopped, and immediately felt something give way in his left calf which was accompanied and followed by stinging pain. He continued to walk about for two weeks, during which the limb grew worse, and he was finally obliged to resort to crutches and without allowing his foot to touch the floor. At the junction of the lower and middle third of the leg, on the inner aspect, there was a depression of the skin which with the muscular tissue beneath had a hardened cicatricial feeling, and the whole of the inner belly of the gastrocnemius was rigid, indurated, and painful. The limb could not be fully extended on account of painful stretching of the tendons of origin of the gastrocnemii. Extension could be slowly made to about four-fifths normal. The limb had been showered with cold water and improperly bandaged. Massage was painful at the first application, but at the second the limb was greatly comforted, and immediately after, the patient could extend it himself without discomfort. In correspondence with this, the rigid, swollen inner belly of the gastrocnemius had become more supple and less sensitive, and œdema had disappeared. Resistive motion was then begun and well executed. At the end of two weeks he could walk very comfortably without support after six massages, and a uniformly applied bandage constantly worn. In seven days more he could go up and down stairs. In twenty-four days he was getting well too fast, as was shown by his having overwalked and relapsed slightly with the former symptoms, but these were very promptly relieved by massage, and the reapplication of the bandage. Seventeen days later, all disagreeable symptoms had vanished, and the leg has been well ever since, now five years, and there has been unusual freedom from the aching stiff feelings that generally linger a long time in these cases.

My friend, Dr. Geo. J. Bull, has kindly reported to me the following case of rupture of the anterior fibres of the trapezius from reflex action in the effort of the patient to save herself from falling: Mrs. C., aged 45, a robust and muscular American lady, weighing one hundred and sixty-five pounds, fell heavily on a slippery side-walk on March 13th, 1875. She did not bruise any particular part of her body, but made a violent effort to save herself, and was badly shaken. Her menses came on prematurely that day;[1] but no other inconvenience was complained of, except a feeling as if she had taken cold in the neck. Next day there was decided pain in the neck and shoulder, which was not relieved by the usual household liniments, but grew steadily worse until I saw her on March 17th, four days after the fall. I found her carefully supporting the left elbow and hand, and inclining her head to the left side. When perfectly still, she felt only a dull ache ; but the slightest movement of the head or arm caused her to cry out with pain in the left shoulder, neck, and back of the head. The pain was distinctly cramp-like, and more severe than anything she had ever suffered, even in childbirth. The skin was not at all bruised, nor was there any sign of injury in the shoulder-joint. There was only great tenderness above the scapula, covering an area as large as one's hand. After half an hour's massage of the muscles of the neck and shoulder, the tenderness was found to be confined to a spot on the side of the neck, a couple of inches above the scapula, where the slightest pressure caused acute pain and spasmodic contraction of the anterior fibres of the trapezius, running up to the occiput. Light rubbing was soon tolerated over the spot, and the pain no longer radiated to the occiput and shoulder. Before leaving, I was able to move the arm freely in all directions without causing pain, and the patient could move her arm and her head a little, while my hand supported the trapezius, without the spasm coming on.

The improvement continued, and the patient was tolerably comfortable lying on her right side. The spasm occurred occasionally in the night, and half a dozen times next day. There was thought to be a slight depression at the seat of the pain, and a swelling behind it. On the eighteenth, after massage, the head could be raised and nodded, and moved to the right without spasm, but any attempt to move it to the left was followed by spasm, and could not be persevered in. Massage was used by me for several days ; voluntary motion of the trapezius was encouraged, and on the 31st of March all symptoms had disap-

---

[1] A patient of mine, æt. 42, convalescing from rheumatism, fell heavily on her nates, and the menses which had been absent for three months came on again.

peared. At present, August 14th, 1875, if the lady sweeps much or takes cold, she has some pain in the old spot, but rest soon relieves it. The doctor has added: "I have not given my notes in full, as they contain little else of interest; but I have said enough to show how a lady, not in the least nervous or hysterical, was relieved by massage of a very severe pain, and restored to a healthy condition more speedily than could have been accomplished by any other therapeutic means."

Of thirty-three cases of myositis of spontaneous and traumatic origin, treated mainly with massage by Johnson (*loc. cit.*), eighteen recovered, fourteen improved, and one was without result.

After complete rupture of muscles or tendons, massage and movements will often be able to educate the remaining well muscles to supplement the loss of the injured one, as the following case would seem to show:

Mr. J. was a moderately corpulent, hale and hearty old gentleman, about seventy years of age. Nearly three years after rupturing the tendon of his left rectus femoris he called to see me, and I gathered from him the following history relative to his case: In getting off a horse-car, thinking it had stopped when it had not, he fell down, and, as it was with difficulty he could stand when he got up, he was taken home in a carriage. The tissues above and around the left knee-joint were swollen and ecchymosed. As he bore with impatience the restraint of rest which his physician put him under, at the end of a week he attempted to walk, but falling down two or three times somewhat deterred him from further attempts at this means of locomotion. Immediately following this came a *merciful* (?) illness of gastric origin, which kept him in bed for a couple of weeks. At the end of this time the extravasation had been absorbed, leaving a depression above the patella, into which two fingers could be placed. Since then he had only been able to walk short distances, two or three squares or so, very awkwardly, and with his mind concentrated on the lame leg, for when he was off guard, which was quite frequently, he would fall.

About this time he consulted me, and after three applications

of massage he walked from his house around the Common and
Public Garden — about a mile and a half.   The following week
he had three more séances, when he again tried his pedestrian
powers more than his business gave him any occasion to do, and
this time extended his walk considerably beyond that of the
previous week, " with greater ease and comfort," he said, " than
he ever believed he would, and without scarcely thinking of the
lame leg."   After three more *massages* he went out of town to
look at a country residence for the summer, and, being seated
in a rear car, he thought they had not quite reached the depôt
he wished to get out at.   But when they started he discovered
his mistake, and pulled the bell-rope, got out when the cars
stopped, and walked back to the depôt on the track, a distance
of over a mile.   The carriage he had engaged was by that time
gone, and none could be had, for it was only a small country
station.   So off he started on foot, and walked from and to the
depôt, over a hilly country, in all about five or six miles, enough
to tire almost any one, he thought, and yet the lame leg felt no
worse than the other next day.   He continued to have such
good use of the limb without any more trouble for four months
longer, when he was taken with a severe illness, which kept him
prostrate for a long time.

The treatment was friction and kneading of the whole limb
with the palm and fingers, varied with acto-passive or resistive
movements of the foot, leg, and thigh in all their natural directions.   While using the acto-passive motion, it was interesting to
observe from time to time the increasing strength and ease of
movement which he was gaining.   That such treatment often improves the nutrition and contractility, extensibility, and comfort
of muscles, and consequently makes them more ready to obey the
mandates of the will, or the automatic action of its subordinate
centres, has been recognized from the most ancient times.

Dupuytren's finger contraction, consisting of hyperplasia and induration of the palmar fascia, the result of slight injury and
reflex nervous irritation, has been successfully treated by means
of massage.   At the meeting of the New York Academy of Medicine, held April 17th, 1884, Dr. Sayre said that by this means

alone he had secured excellent results in two cases without resorting to any operative procedure whatever. On account of the
atrophic condition of the areolar tissue, he believed that massage
and similar manipulations were of so much service in this affection. Von Mosengeil narrates the case of a violinist who contracted the affection in consequence of an injury to the palm of
the hand by a fiddle peg. The ring finger was particularly affected.[1] During the massage, the contracted fingers were extended.
In this manner, the trouble was completely and permanently removed.

In 1874, the venerable Dr. Henry I. Bowditch sent me a well-
marked case of *elephantiasis* in a girl fourteen years of age. She
looked pale and delicate, but enjoyed good health, notwithstanding
the fact that the affected leg, from the toes to the hip, had the
appearance of belonging to a Cardiff giant, and certainly represented at least one-half of her total weight. Half an hour of arduous,
deep massage caused a decrease in the circumference of the calf of
half an inch at a definite point. I never saw nor heard from the
patient again. She had been to an irregular practitioner one hundred times for electricity, under promise of being cured, and it is
quite likely that her parents took her back to him.

I was more successful with my next case, which was sent to
me by the late Dr. George H. Gay in November, 1877. The patient, sixty-nine years of age, presented the symptoms of *elephantiasis*, but was otherwise well. The history of his trouble began
four months before he came to me, by his being seized with sudden and severe pain in the middle of the lower part of his back,
accompanied by chills, while the patient was at the water-closet
in undress, having gone there soon after being circumcised. He
could not get back to bed without assistance. In the course of
an hour the pain extended to the left hip and down the left leg,
and soon the leg became swollen. For three months and a half
afterwards, there was loss of power in defecation, but the bladder
was normal. When I first saw him, the leg, from the toes to the
knee, was of a dusky, bilious color, looked like a side of bacon,

---

[1] In eighty-eight cases out of one hundred and five, Dr. Keen found the ring finger
affected.

and felt almost as hard as a board, so that it was with difficulty finger pressure would leave an indentation. The skin, superficial fascia, and connective tissue were hypertrophied and indurated, with here and there on the surface patches of eczema. To the patient the leg felt numb, as might have been expected from so much pressure upon sensitive nerves. He could walk a short distance naturally. The calf at its greatest circumference was two and three-quarter inches larger than the other. Daily at first, and every other day afterwards, I used showering, alternately with hot and cold water, and this was followed by deep, firm, upward kneading, which was begun at the uppermost part of the infiltration, and gradually proceeded with downwards; and the decrease in circumference obtained in this manner was held and continued by two or three layers of tightly-applied bandages and instruction to keep the limb elevated. In four days the calf had decreased one and three-eighths inches in circumference, the leg was supple and more comfortable, the capillary circulation was better, and the patient could walk easier. At the end of twenty-two days, there was very little of the bacon, board-like condition left, and massage made greater impression, twenty minutes of it causing a temporary decrease of one-quarter of an inch in circumference, whereas at first but one-eighth of an inch was gained; and the total diminution was two and nine-sixteenths inches, three-sixteenths of an inch more than the other calf. The patient had twenty sittings in five weeks, at the end of which the calf was the same size as the well one, the skin could be pinched up in natural folds, but the ankle was still somewhat larger and harder than the other. He was then discharged, with an elastic stocking, and advised to continue the showering and keep the limb elevated as much as possible. A month later, his exercise consisted in a daily three-mile walk, and that without discomfort. Six months later, he had laid aside his elastic stocking, and the leg continued well.

The day on which I finished treatment of the above case I found in Schmidt's *Jahrbücher*, No. 1, Bd. 173, 1877, a brief account of two cases of elephantiasis, treated by manipulation and bandages, by Prof. Von Mosengeil, of Bonn. In one, both legs were effected from the toes to the knees. At first massage alone was used with considerable improvement; then firm compression, by means of roller bandages, was added. After five or six weeks, the patient could walk better, and the skin was thinner and more movable. The other case was less severely affected, and greater improvement resulted.

When œdema is dependent upon disease of the heart, liver, or kidneys, or an altered condition of the blood, massage will not act in a curative manner, though temporary alleviation may be gained. But œdema continuing when the cause has been removed is generally speedily absorbed under the impetus of massage upon the lymphatics and capillaries. In recent thrombosis, massage might be dangerous, and convert the case into one of embolism. Sir W. Fergusson adopted this plan with advantage in two cases of aneurism of the right subclavian artery. He first emptied the sac by pressure with his thumb, then squeezed and rubbed the opposing surfaces against each other, so as to force the fibrin into the artery. The effect was immediate and striking. In the first case there was giddiness, in the second partial hemiplegia, so that in all probability some of the fibrin had been carried into the vessels of the brain. In the first, after one or two repetitions of the manipulation, all circulation in the vessel and its branches was arrested, and the aneurism became smaller and firmer. After unusual exertion, it burst into the brachial plexus, and the patient died seven months after the first manipulation. In the other case, the tumor became gradually less, and the patient was alive and well two years afterward. (Druitt.) By the same procedure, Herbert Page, of Carlisle, also treated successfully a case of popliteal aneurism.

The œdema and superabundance of callus sometimes met with after the union of fractures can both be lessened or removed by the use of massage, at the same time that the circulation is increased and the muscles restored to activity. Passive and active movements should also be encouraged.

In a case of myxœdema of four years' duration occurring in a man 56 yrs. of age, who was sent to me by Prof. R. T. Edes, the result of a course of massage every other day for six weeks was most satisfactory.   The boggy, brawny, waxy condition of the tissues gave place to suppleness, freedom of motion, improved mental tone, and better hearing.   Four years have elapsed and the improvement has continued.   The man is practically well and attends to his business.

CHAPTER XIV.

THE TREATMENT OF SCOLIOSIS BY MEANS OF MASSAGE.

" We'll come dress you straight."

THE results obtained from the employment of massage in carefully selected cases of lateral curvature of the spine seem to have been so successful that they deserve more than a passing notice. Dr. Landerer's paper on this subject, together with the discussion of the same at the Congress of the German Society for Surgeons,[1] are worthy of our consideration. Our author regards habitual scoliosis as that arising from superincumbent weight as genu valgum sometimes does, the weight pressing bones and joints in wrong directions and ultimately causing anomalous growth. It is in the earlier stages of this form of scoliosis that he has found massage to speedily bring about recovery, and in the later stages where the deformity has become fixed, intercostal neuralgia and painful tension of the muscles are relieved, and the patient made comfortable by the same means. This form of scoliosis is to be kept separate from the static, rheumatic, traumatic, empyematic and other kinds, in which it would be well to include that arising from disturbance in the central nervous system.

The production of the natural curves of the spinal column is clearly explained. In early childhood the spinal column is straight. The normal S curve arises from the combined effect of gravity and muscular action, the former alone would cause a simple backward curve, a total cyphosis; the latter modifies this and forces it into a serpentine curve. The action of both is to shorten the spinal column. While our observation would agree with that of the

---

[1] Deutsche Zeitschrift f. Chirurgie. Band XXIII, Heft. 5 und 6.
Deutsche Med. Zeitung, No. 32, p. 539.

author, that marked serpentine curves, especially deep lordoses in the lumbar regions are frequently found in those of great muscular strength and in stout people of medium stature, we would beg to differ from him in his statement that those who are tall and slim spare their muscles by throwing the centre of gravity of the upper portion of their body as far back as possible. More often the latter stoop or are round-shouldered, and when they maintain an erect attitude the absence of marked curves may be owing to the muscles not being sufficiently strong to curve and shorten the spinal column.

The upper and lower extremities of the cervical portion of the spinal column are approximated by means of the muscles at the back of the neck, the contraction of which changes the former convex backward curve of infancy to a concavity. This result is aided by the effort to maintain the centre of gravity, for the middle and lower parts of the cervical region carry the most of the weight of the chest. The thoracic organs and even part of the weight of the abdominal organs are suspended from the first and second ribs and from the region of the sternum to which these are attached, and these again are held by means of the scaleni muscles and by them raised during inspiration, so that the weight of the thorax is transferred to the middle and lower cervical vertebræ where these muscles are attached.

As the dorsal region of the spinal column has but little strain upon it in either direction, it remains convex posteriorly as in infancy. But it is otherwise with the lumbar region which becomes convex anteriorly owing to the action of the large muscles on its posterior aspect, which changes the previous backward convexity into a concavity. The lumbar region carries the major part of the weight of the intestines; it is here that the mesentery is attached and also the psoas muscles. These muscles, when the thighs are fixed as in standing and still more in assuming the erect position, make a downward pull upon the lumbar vertebræ in the same manner as the scaleni muscles do upon the cervical portion, thus necessitating a strong counterbalancing action from the muscles on the posterior aspect.

The explanation of lateral curvature is not so easy. Slight

lateral deviations are frequently found in otherwise well formed people, and according to our author it is not agreed whether this should be called physiological scoliosis or not. We would at once anticipate his conclusion by inferring that this gives no trouble so long as the muscles are strong and active, and that therefore measures to prevent or restore their strength and activity should be employed. Amongst classical statues in the Louvre and British Museum Dr. Landerer has not been able to find any examples of the physiological scoliosis. The spinal column being freely balanced perpendicularly upon the pelvis, and thus held by the muscles on each side of it, as a freely balanced mast would be by ropes, it follows that when deviation occurred to one side or to the other it must be on account of muscular relaxation, as the mast would deviate if one of the ropes were slack. Increased weight in the perpendicular direction alone does not cause lateral deviation. The effect of gravity upon superincumbent or suspended weight will not produce scoliosis so long as the muscular structure is normal.

In Swabia, where the home of our author is, he tells us that women and girls carry heavy loads upon their heads up high mountains, but in spite of this scoliosis amongst the laboring classes is seldom met with. On the contrary, very prettily formed figures are almost exclusively found, due in great part to this exercise. The maidens of Capri also carry heavy weights upon their heads and are remarkable for their faultless development. The muscles are thus made powerful to oppose strong lateral resistence.

In incipient scoliosis the spinal column is exceedingly flexible and this great mobility should be regarded as the first sign of lateral curvature, especially when accompanied with flat back.

Autopsy of scoliotics reveals atrophy and fatty degeneration of the muscles of the back, especially of the concave side. We would have supposed that it would be greater upon the convex side where the relaxation would be.

We are all more or less critical in observing any slight obliquity of the shoulders and lateral deviation of the spinal col-

umn, but somewhat indifferent toward the antero-posterior
direction of the median curve. Only in the most recent ladies'
fashions does our author find that a well developed median
curve is necessary and that a deep concavity in the lumbar
region—lordosis—is pretty. The latter is formed artificially by
the bustle or *tournure*.

In health the antagonistic muscles keep the vertebræ at proper distances from each other. If there be muscular weakness
they will sink down upon one another, and in sitting or standing if the muscles act obliquely the spinal column will deviate from
its natural position and the vertebræ will be pressed together
causing disturbances of their circulation and nutrition until they
finally become deformed. In brief, for the *preservation of the
spinal column in a natural position, healthy muscles are necessary. Habitual scoliosis arises from superincumbent weight, the
original cause of which is weakness of the muscles, and therefore the treatment has to be directed to them.*

The results of treatment proved to our author that his views
were correct. Massage goes further than gymnastics, and what
these accomplish slowly massage does in a direct manner by
the hands of the surgeon. With cases of scoliosis in the first
stage which permanently improved in a few months, the children felt stronger and steadier in the back and held themselves
more erectly even after a few *séances*. Though the immediate
effect of massage was quite evident, yet part of the improvement
disappeared within a short time, but the gain gradually became
lasting. The method employed by Dr. Landerer is the following:
The child is laid upon the abdomen, the trunk bare to the lower
half of the crests of the ilia, the arms stretched forwards. The
extensors on both sides of the back are percussed with the balls
of the little fingers from their origin on the pelvis upwards to
the neck; first gently, then more vigorously. The concave side
is percussed more strongly than the convex. The muscles on
the side of the trunk, so far as they are connected with the
spinal column, come in for a share of the same. Then the extensors of the trunk are stroked with the fingers held in a perpendicular manner.

We do not see why percussion should be used more energetically on the concave side, unless it be carried to an extreme degree so as to tire out the contracted muscles and thus cause them to relax. Percussion has much the same effect as faradization, and can be used in moderation to stimulate muscular contractility. Our method of proceeding in such cases differs somewhat from our author's, for seeing that lateral curvature may not only be favored, but actually produced in those predisposed thereto by persistently sleeping on one side with high pillows under the head, a great part of the treatment should consist in having the patient lie upon the opposite side so as to reverse the curves. And it is better that most of the massage should be administered with the patient in this position, for massage helps to relax contracted muscles when they are stretched, and it stimulates the contractility of relaxed muscles. The insertion of muscles implies their attachment to the more movable parts, and as their returning circulation almost always follows the course from their insertion to their origin, it would seem much better to proceed with the massage from the neck to the pelvis. Deep manipulation, rapid pinching and rolling of the muscles have much greater influence in stimulating their nutrition and contractility than the stroking of our author which acts more upon the superficial circulation.

Dr. Landerer next uses manœuvres which act more especially upon the skeleton, the spinal column and the framework of the chest—the *redressement*. These resemble the rotation and torsion movements which have hitherto played an important rôle in the treatment of scoliosis. After this the spinous processes are acted upon by stroking from behind upwards, and by pushing them directly towards the concave side so as to equalize the curves. The immediate effect of all this is quite perceptible. Then the depressed parts of the thorax are raised by one hand gliding from the anterior aspect of the chest backwards, raising the concave side, whilst the other presses down the projecting parts on the convex side by stroking with pressure from the spinal column forwards around the chest, and thus literally remodelling the youngster. Prominences elsewhere, as under the scapula, receive special pressure. Sometimes these operations can be done better while the patient stands or bends forward. Precise rules cannot be given, as no case of scoliosis exactly resembles another. A well-schooled anatomical eye, therapeutical instinct, inclination to treat such cases, together with experience, will lead to accurate treatment and good results.

For special exercise of the extensors of the back the patient is placed and fixed upon the anterior aspect of the legs on a table, the body projecting beyond the edge and sufficiently supported. Then the patient elevates the trunk from the horizontal position backwards to a right angle, if possible, by energetic contraction of the extensors. We think the patient should often be assisted in elevating the trunk.

To wind up the *séance* Landerer allows the patients to suspend themselves five or ten times by means of Sayre's apparatus, the hand corresponding to the higher shoulder should then be the lower. When treatment has been continued for a few weeks active exercise is allowed by means of a basket carried upon the head with a weight of three to five kilos in it, which is supported by the hand of the depressed side.

Whilst the patient is suspended we have found it advantageous to manipulate, roll and percuss the muscles of the back; well known beneficial aids, not spoken of by our author, are rowing, sitting on an inclined plane with the higher shoulder towards the higher end so that the patient must bend in this direction and make a constant effort to keep from sliding off. Placing the arm of the side that corresponds to the convex dorsal curve across the front of the chest in an upward direction so as to relax the serratus magnus and rhomboidei muscles, whilst the arm of the opposite side is placed obliquely downwards across the back so as to make the same muscles on this side tense, forms a useful exercise by literally unscrewing the patient. With the arms in like position, the patient sitting on an inclined plane can pull elastic bands and get still more effect as recommended by Professor Sayre.

This treatment is usually given once daily, but in difficult cases or where a speedy result is desired, it may be used twice a day.

Our author gives details of the treatment and results in eighteen cases. Seven of these he places little importance upon, as they were mild and would probably have recovered perfectly, if not so soon, under the usual orthopædic exercises. But it was quite otherwise with five of the cases which were much worse. In these treatment was begun without any hope of improvement and only at the request of the patients. In from twelve days to five weeks there was marked improvement, they were soon able to do without their supports, the deviation in the spinal column decreased, the shoulders became more horizontal and pain and tension disappeared. Of the remaining six cases four were improved and two got well. At the time of writing several were under treatment so that the results may be still more favorable. At a time of life when the skeleton was so consolidated as it was in the case of an eighteen-year-old girl, it was not thought possible to produce any remodelling, but after two months' treatment it was hardly possible to perceive any deformity. Landerer finds an analogy to this last case in several of severe so-called inflammatory flat-foot or *tarsalgia adolescentium* in which he succeeded in obtaining "perfect reformation" of the foot by means of massage of the plantar surface and of the leg, even at twenty years of age.

We think that another analogy can be found in the relief that massage affords in cases of rheumatic gout. It relieves the tenseness of the soft tissues and makes them more supple, so that they adapt themselves much easier to the fixed nodosities.

Our author believes in the use of supports and corsets for alleviating and correcting the position of advanced scoliosis; but when there is any prospect of improvement he considers them contraindicated, for the little work that the muscles of the back may be capable of performing is taken away by supports and atrophy speedily results so that the muscles can no longer be used. After two or three massages he found that most of his patients could do without supports and rejoiced once more in free and lively motion.

In the discussion which followed, Herr Volkmann deprecated the wearing of a plaster jacket or felt corset by day and night. He has the corset removed at night, and in the morning the patient is bathed, douched and *masséed*, and practices movements, as advised by Sayre, and about 11 o'clock the corset is again applied.

Herr Loebker stated that he believed in the energetic treatment by means of massage of those muscles which contribute to the support of the spinal column. He does not apply any supports, and improvement takes place from the time these are laid aside.

Herr König, of Göttingen, said that Landerer's method was in advance of that hitherto employed by him; yet he would not renounce altogether the corset treatment in favor of massage. He would limit the use of the corset to school time. Experience had taught him that the complete removal of the corset all at once hinders improvement.

Herr Kölliker, of Wurzburg, remarked that the most essential difference between Landerer's treatment of scoliosis from that hitherto employed consisted in energetic percussion of the muscles. He mentioned a very severe case of scoliosis with three curves which he had treated daily for three months by means of massage and percussion for several minutes night and morning, and thereby obtained a brilliant result never before equalled in his experience. With scoliosis of the second degree the corset should be applied in the intervals between massage.

CHAPTER XV.

WITH A REPORT OF ITS RESULTS IN OVER SEVEN HUNDRED CASES : RECOVERY
IN ONE-THIRD OF THE USUAL TIME UNDER OTHER METHODS OF TREAT-
MENT.

"Oh, this is well ; he rubs the vein of him."

IN various affections of joints, more than of any other parts of
the body, massage has been used successfully. Testimony of this
is so abundant that it will be impossible to make use of it all here.
Those joints whose capsules are accessible to the immediate pressure of massage have responded most favorably to such treatment,
while the hip and shoulder joints that are covered by muscles and
not so easily got at have not yielded so readily. For convenience,
the effects of massage upon joints may be spoken of as local, revulsive, and sympathetic. It may be used to increase the circulation in and around a joint, or to squeeze congestion and exudation
out of it with one hand, while the other pushes along the circulation in the veins and lymphatics above the joint, or the joint may
be avoided altogether while the muscles above and below it are
masséed, thus making more blood go through them and less to the
joint. It would seem as if the vascular and nervous supplies of
joints had been pre-arranged with special facilities for the influence
of massage upon them. With regard to the direct, sympathetic
and reflex effect of massage upon the nervous supply of joints, the
broad generalization so well expressed by Dr. John Hilton is of the
greatest interest, namely, that "the same trunks of nerves whose
branches supply the groups of muscles moving a joint, furnish
also a distribution of nerves to the skin over the insertion of the
same muscles, and the interior of the joint receives its nerves

256

from the same source. This implies an accurate and consentaneous physiological harmony in these various co-operating structures. Without this normal consentaneous muscular and sensitive
function, precision of action would be lost, and unnecessary exercise of muscular force would be employed during the performance
of any of their functions."

Still another interesting fact concerns us at the present time,
and that is the abundance of Pacinian bodies around the joints,
whatever significance this may have. When cramp can be localized, the pain is mostly felt at the joints, and where in tissues accessible to massage pain can be severely felt, there massage is most
efficacious and agreeable.

Prof. von Mosengeil, of Bonn, has made some very interesting
and important experiments by injecting the cavities of corresponding joints of rabbits with indian-ink, and in this way he has
demonstrated that resorption takes place from these cavities by
means of lymph spaces and stomata communicating with lymphatic
vessels, and through these with lymphatic glands. With each
rabbit he *masséed* one of the joints, and left the corresponding
joint untouched. The swelling that arose from the injection always disappeared rapidly under massage, and upon examination
of the *masséed* joint it was found emptied for the most part of its
colored contents. Even when the examination was made shortly
after the injection and the use of massage, there was scarcely any
ink found in the joint; part of it was found upon the synovial
membrane, and upon microscopic examination it was seen that the
greatest part had been forced into and had penetrated through the
synovial membranes, and the darkened lymphatics could even be
seen with the unaided eye from the injected joint to the lymphatic glands, and these latter were black from the absorption of
ink. Upon examination of the joint cavities that had not been
masséed, the ink was still found in the joint mixed with the syno-
via in a smeary mass, and it had not even penetrated into the tissue of the synovial membrane. With the removal of the effusion
by the use of massage, the stiffness of the joints was improved,
and the same appearances in the lymphatic vessels were always
obtained. This evidence was highly confirmatory that the struc-

ture and functions of the synovial membrane are similar to those of the pleura and peritoneum where the pump-like action of respiration causes the lymphatics to suck up and propel onwards both natural and morbid products. The sheaths of tendons having a similar structure, would be influenced in a like manner by massage, passive and active motion.

Joints, tender and swollen, that do not admit of massage being applied directly upon them, may be approached by commencing on the healthy tissues some distance above them and nearer the trunk, with gentle stroking in the direction of the returning currents of the circulation, and gradually proceeding downwards. The healthy tissues beyond the seat of the malady should also be similarly treated, as the circulation is hindered in getting to and from them. Besides the soothing effect of this which enables one to gradually encroach upon painful tissues, the circulation is pushed along more quickly, so that exudations are carried off more easily. After working a few minutes in this manner, deep manipulation may be brought into play, proceeding in the same direction by beginning above the painful joint and making the greatest pressure upwards, while gradually approaching the objective point, the effect of which is to lessen pain without decreasing ordinary sensation. As near as we can get to this by alternate stroking and kneading, we will make a review on both the proximal and distal sides of it, and by repeated efforts of this kind in the course of fifteen or twenty minutes, we will usually be able to make gentle, firm pressure upon the sore joint, and this pressure can imperceptibly have motion added to it, thus constituting massage properly so-called, by which effusions and exudations are spread over greater surface, pressed through the meshes of areolar tissue as water is out of a sponge, and brought into more numerous points of contact with veins and lymphatics, and these are very materially aided in their resorptive functions by the pressure of the massage. It should not be forgotten that, when a light touch is disagreeable, firm pressure often affords relief. Recent periarticular effusions that have not become organized are thus speedily dispersed and absorbed, while superabundance of intracapsular fluid is pressed into the absorbents, the function of which, within the joint, is increased by the pressure from without, and by the acceleration of their current from the massage above the joint.   In recent sprains and synovitis this method is rational.   It relieves the pain by removing pressure from terminal nerve-filaments; it reduces elevated temperature by hastening absorption, and thus removes the tension which causes lymphatic and venous stasis and exudation, and at the same time it increases the area and speed of the circulation.   The comfort to a joint even after a single sitting would hardly be believed unless felt or witnessed. The effect is enhanced by a bandage well applied after the massage, but the pressure of a bandage, though it affords sup- port, is at the same time by its continuous pressure a hindrance to the circulation and will not take the place of massage, which is an intermittent pressure and an aid to the circulation.[1]   These procedures may be done once or twice daily, or every other day, as may seem necessary.   Patients are generally ready to recognize improved power of motion and to make use of it.   If they do not, the operator can judge for himself by passive and resistive move- ments how much or how little the joint is capable of, and encour- age or restrain motion accordingly.   In recent cases it is preferable that the patient manifest the first desire to make use of the joint, for it is often impossible to tell how severe the affection may be, and if the patient be not of a nervous temperament, sensations and inclinations can generally be trusted.   The vast number of sprains of all degrees of severity that have recovered in from seven to nine days under massage would seem to prove either that rupture of ligaments, tendons, and laceration of muscles, with effusion of blood into the joint, occur much less frequently than is supposed, or else are of much less serious import when treated by massage. Further advantages of massage in these cases are the prevention of the formation of adhesions and the loosening of old ones that are neither too deep nor too strong.   After time for repair has elapsed, in order to gradually increase the strength of the muscles,

---

[1] Kraske has demonstrated that the application of a rubber bandage to the leg of a rabbit for six hours has produced vitreous degeneration of the muscles. *Archiv für Chi-rurgie*, 1879, 1881.

as well as the confidence of the patient to use them, there is nothing better than resistive motion, alternately resisting flexion and extension or other natural movements of the affected joint while keeping the resistance less than the strength of the contracting muscles, so that the patient may not recognize any weakness.

When joints are lax and muscles flabby, in the absence of acute symptoms, vigorous deep manipulation and percussion with brisk, active, and resistive movements, followed by a tight bandage, are indicated, but passive motion pushed until there is a feeling of resistance should be avoided. With capsular and periarticular thickening, induration and hyperplasia of an indolent character, kneading with one hand upon the affected structures, and stroking with the other above them, will play the most important rôle, while increasing passive motion will be persisted in. In such instances, the leaving off of a bandage and encouraging active motion can generally be done with safety, and will be regarded by the patient as marked evidence of improvement. As motion is impaired in all sorts of joint affections, it is well that the muscles on each side of them should be stimulated by massage.

Massage disintegrates newly formed granulation tissue, removes the stasis which it has occasioned, presses the white corpuscles and transuded plasma into the lymph current, at the same time the newly formed capillaries that feed this granulation tissue are ruptured and undergo retrograde metamorphosis as well as the crushed mass, and thus the formation of connective tissue and the subsequent change of this into cicatricial tissue, which often causes pernicious retraction, is prevented or limited. Hyperplastic tissue firmly organized, solid like India-rubber, and not sensitive, is probably non-vascular, owing to its pressure upon and obliteration of the capillaries which previously nourished it. Upon such tissue it is hardly possible to accomplish anything by massage, though it is considered by Gottlieb, Billroth, and others that by vigorous perseverance impervious blood and lymph vessels may be dilated, and resorption promoted. Certainly the fibrous thickening of the capsule sometimes met with in chronic serous synovitis, which involves the synovial membrane and peri-synovial cellular tissue, must yield but slowly to massage, and still more slowly to time alone. While in hyperplastic synovitis with connective tissue thickening of the capsule, we would use massage with energy, for the same reason that in trachoma of the conjunctiva various irritants are used to induce congestion, and thereby cause a retrograde metamorphosis of the sclerosed tissue; in other more acute, sensitive conditions, such as recent sprains and synovitis, massage would be used with great care and gentleness. In the former case, it is used as an irritant, to create a slight inflammation; in the latter, as a sedative and antiphlogistic in the manner already described. In place of the dry, disconnected details of pathological anatomy that rack a student's brain, we hope that some one will attempt to show more clearly the analogies betwixt similar morbid processes in different parts of the body. In trachoma of the conjunctiva, pathological changes occur similar to those found in hyperplastic synovitis. What an advantage it is, by turning up the eyelid, to be able to see unchecked granulations extending into the stroma of the conjunctiva which later contracts, atrophies, and changes into fibrous cicatricial tissue with obliteration of the blood-vessels.

In highly acute arthritis of any kind, attended with fever, massage would not be thought appropriate until the disease had assumed a subacute or chronic form, and the fever had abated. Then massage might be used with benefit, provided there was no solution of continuity, true anchylosis, or risk of hastening absorption of inflammatory products pernicious to the system. In disease of bone or cartilage massage would be useless. Further classification is unnecessary, as each case must be judged by itself.

A sprain may be defined as a partial and sudden displacement of two joint surfaces, followed by immediate replacement; and if the patient has fallen from a height there will probably be contusion of the articular surfaces and soft tissues as well. The attachments of the joint on one side are stretched beyond their natural limit, on the opposite side unduly compressed. But if the patient has fallen squarely on a joint, there may be only contusion of its surfaces, without strain of its attachments, and in this event the external symptoms are slight or absent, while the discomfort may be great.

The higher an ankle joint is from the sole of the foot, the greater its liability to sprain, and if any one else has made this observation I am not aware of it. People with high ankles, when well, will be surprised to be told of this peculiarity which has caused most of them to suffer.

In 1877, I published an article in the New York *Medical Record*, giving the results of massage in the treatment of 308 cases of sprains, joint contusions, and distortions, by seven independent observers, who were French, German, and Scandinavian surgeons, besides a few cases of my own, to illustrate certain other effects of this treatment. In these cases, which seemed to be of all degrees of severity, the average length of time for recovery was found to be 9.1 days, and this time would have been much less if the 39 cases had been omitted in which massage was not begun until from ten days to three months after the injury, in many of which it is stated other methods of treatment had failed, and which required, on the average, three weeks of massage before recovery. A study of 55 cases treated in the usual manner by these same observers, showed the average time for recovery to be 26.16 days, or nearly three times as long as similar cases treated by massage. The earlier after the injury manipulation was employed, the sooner recovery followed. The advantages of massage in such cases would seem to be more speedy relief from pain and swelling, and earlier and more perfect use of the injured joints than are usually obtained by any other method.

With regard to sprains, Estradère thus expressed himself in his work on massage, published in 1863 : "I have always been of the opinion that massage ought to be avoided whenever there were indications of inflammation, *but this is not so with reference to sprains.* Indeed, according to the opinion of MM. Bazin, Bonnet, Brulet, Elleaume, Girard, Lebatard, Magne, Méry, Quesnoy, Ribes, Rizet, etc., who have recently published their observations upon sprains cured by manipulation, such affections ought to be treated immediately by this procedure. The pain, ecchymosis, and swelling disappear as if by magic. Others, pushing the same thing a little farther, make use of massage even when a laceration of a malleolus exists, persuaded that they have made one step towards cure by removing the pain and swelling, and by replacing luxated tendons. I admit that if the distinguished physicians who have related these facts had not persuaded me by their perfect accord in this matter of the harmlessness of massage in such cases, I should have been very guarded in venturing to suggest it; but the facts are indisputable; however strange they may seem, we must admit them. As Rizet says in his monograph on the treatment of sprains by massage, impressed by the words of Baudens at the Academy of Sciences, that of 78 amputations of the leg, 60 had their origin in sprains, it was with eagerness we seized opportunities to try this means, which, far from disappointing us, has given us unexpected success. . . . At the present time those physicians who have it in their power to make use of this method are unwilling to make trial of it, but allow their patients to go to bone-setters, charlatans, fortune-tellers, and sprain-blowers, who accompany their manipulations with divers mysterious signs. But we ought not, says Nélaton, to reject a useful means because it has been used by those unskilled in the medical art."

Dr. Beranger-Feraud, an old army-surgeon, gives an account of four hundred sprains which he treated successfully by means of massage, in *L' Union Médicale du Canada* (*Philadelphia Medical Times*, November 20th, 1880). These are spoken of as slight, medium, intense, and complicated, and the conclusion is that the nearer the massage is used to the time of the accident, the sooner is the recovery; and a sitting ought to last until all feelings of distress and pain have disappeared.

Dr. Gassner, of Würzburg, military surgeon, has reported twenty-four cases of acute serous joint-inflammation from sprains, treated successfully by means of massage.[1] The average stay of the patients in the hospital until they were perfectly well was 8⅓ days, while of 13 cases of similar injury, the average time in the hospital under rest and immobility was 28 days. Only in severe injury

---

[1] Allgemeine Med. Central-Zeitung, Sept. 4th, 1875.

to the joints with considerable rupture of capsules, ligaments, or muscles, or harm to the bones, would he make use of immovable dressings, and after repair massage would then be used to get rid of the consequences of previous immobility. Well he remembers the tardiness of recovery of his own sprained ankle and contused knee under the many weeks of regular treatment, and much he regrets not then having had massage. He quotes from Billroth, Volkmann, and Erichsen to show that in spite of the most careful and conscientious treatment in the usual manner, even apparently light sprains sometimes lead to serious secondary consequences.

Berghman, of Stockholm, has treated successfully 145 cases of recent traumatic joint affections, contusions and distortions, synovitis with serous effusion, or effusion of blood into the joint-capsule.[1] Seventy cases of these affecting the ankle recovered on an average of 6 days, and twelve sittings; while 38 cases of old sprains required 22 days and forty-four sittings. If a plaster-of-paris dressing had been applied even for a short time, there was no prospect of speedy cure by massage; for under this the connective tissue proliferates and soon acquires a plank-like hardness.

Though this *matted, board-like* condition of the areolar tissue was a marked feature in a case of old sprained ankle which, as a last resort of the therapeutics of despair, had had a plaster-of-paris dressing applied for six weeks, which had just been left off before the patient came to me for massage, yet having had but two massages in one week, the patient began to walk the following week without crutches, and in two weeks walked well. As it was with the utmost difficulty that this patient could have any massage at all, I gave her somewhat lengthy applications, and it was interesting to observe the suppleness of tissue that was manifestly being gained every few minutes. For several weeks at first the sprain was regarded as of no account, but gradually became worse, and for more than a year the patient had been obliged to use crutches.

Mullier treated 37 cases of sprains with massage, and found the mean duration of treatment till recovery to be 9 days, against 25.6

---

[1] Schmidt's Jahrb., Bd. 172, p. 172.

days that 42 cases required under immobilization. As a means of diagnosis, Mullier found massage highly advantageous; for in several cases the presence of fixed pain, of ecchymosis around and over the malleoli, of swelling and functional disturbance, would have led to the opinion that fracture was present; but after a few massages the uninjured continuity of the bone could be recognized, and a proportionately speedy recovery was obtained. In others only after repeated massages was the swelling so diminished as to reveal the existence of fracture, and then appropriate immobilization was applied.[1] When swelling is tedious and difficult of removal by massage, we ought to suspect more serious mischief than sprain alone.

I have attended a number of mild sprains of the ankle where the patient had hobbled about for two or three weeks, in which one or two massages afforded complete and lasting relief from pain and hindrance of motion. But the following certainly was not a mild sprain : Mrs. W. L., an active lady of strong muscles, fifty-six years of age, in descending the stairs of a hotel caught the heel of her boot on one of the brass covers, and did not turn her foot, but turned herself over the foot and fell headlong downstairs. The ankle was severely sprained, and it was impossible for her to walk or even to put her foot upon the floor. I saw her within an hour after the injury before swelling had become great. The capsule of the joint was tensely swollen and projected in front of the external malleolus, and the patient had no control over the foot. The tissues over the joint were extremely sensitive, and could not bear the contact of either cold or hot water. On examination, no evidence of fracture could be found. I expected greater swelling, and a six weeks' stay in the house for the patient, as the injury was certainly of such a nature from its mode of occurrence as to produce all the evils that may be found accompanying a severe a sprain. I *masséed* the limb twice daily, and after each massage applied a bandage tightly, and on the inside of the limb a splint. At the end of a week, the patient could walk a little and in ten days was going about freely on foot. My friends, Drs. Stoddard and Corey, as well as

---

[1] Virchow und Hirsch's Jahresbericht, 1875, Vol. ii., p. 333.

the authors cited above have reported similar cases in detail, with like favorable results.

The following two cases illustrate the aphorism of Hippocrates, which says that "*anatripsis* can bind a joint that is too loose and loosen a joint that is too rigid."

Although Miss C. was a young lady of good muscular vigor and firm tissues, yet, perhaps from presuming too much on these very qualities, she had sprained one of her ankles three times within two years. The last injury was naturally the most serious, the foot turning violently inwards as she alighted on a coil of rope while jumping into a row-boat. For the two or three weeks following she was treated with rest, bandages, etc., and after that she got about on crutches, walking stiffly and with pain, and thus she continued for three months without further improvement. Despairing of anything being done, she reluctantly consented to try massage, for she had had ordinary rubbing every day since her accident, which passed for massage. About three months after her mishap, I was called, and on examination found that there was still considerable effusion in front of the external malleolus and behind the internal, pressure on which excited sharp pain, in the former more than the latter, and to these places was referred the pain which was aggravated by passive motion of the foot, and this pain seemed to be the chief symptom in limiting the passive motion to a very slight degree of flexion and extension. Stiffness and weakness of the muscles from the knee downwards, with induration of the cellular tissue, were also marked features of the case. After twenty minutes malaxation, or kneading with the palm of the hand and fingers, alternating with friction in an upward direction as far as the knee, the effusion was slightly diminished, the tissues were suppler, the limb felt more comfortable, and yielded more readily to passive motion. The patient could now flex and extend the foot herself somewhat, which, before the massage, was almost beyond her power to do. At the second visit I added a little resistence to the voluntary flexion and extension of the foot, but this was almost a make-believe, so feeble were her own efforts in moving it. At the third visit the spots which had been painful on pressure could bear vigorous manipulation very comfortably. Henceforth friction, malaxation, passive and acto-passive motion were persisted in half an hour or so daily, and after five massages the patient walked about the house without crutches or any other aid, and did not require the use of them again. After the sixth *massage* she went up and down stairs naturally, and after the eighth walked half a mile, then eight days from the time this treatment was begun. Four more visits were made on alternate days, and at the last one I tied a handkerchief around the metatarso-phalangeal joints (ball of the foot), and to this attached the hook of a spring-balance, the indicator of which was pulled out to eighteen pounds by flexing the foot (contraction of the anterior muscles), which is a severe enough test, as any one can ascertain if they will take the trouble to try. The patient had had no relapse three years afterwards.

Much of the immobility of the joint in this case seemed to me to be due to moderate tonic spasms of all the muscles of the leg, and I at once succeeded in convincing the patient that it was to a great extent within her power to cultivate the faculty of *voluntarily* relaxing them, so as not to resist the passive motion, which latter was proceeded with gently and tentatively, not forcibly, and with all the leverage afforded by the foot, but being limited by the slightest approach of pain and involuntary resistance, and thus in three or four days there was gained free movement of the joint.

In marked contrast with the condition of tissues observed in the affected limb of the previous case and the effect of massage upon it, is the flabby, atonic state of muscles and laxity of joint in the following case. Miss A. had been in a nervous, dyspeptic, half-invalid, loose-jointed condition for sixteen years. To pressure (I ought say, to touch) either on the spinous processes or on the muscles on both sides of them, her back is fearfully sensitive. High pressure in school-life was said to be the cause of this state of affairs. There was no history of uterine trouble, nor anything of a hysterical nature about the case. For several years she had been growing flat-footed, for which shoes of such a make as would preserve the arch of the feet were advised. Nine months before I saw her she was walking along an uneven road, her foot slipped, and in the effort to regain her balance so as to save her back, she made a misstep, twisting her left foot inwards or outwards, she could not tell which. The foot swelled to about one-half more than its normal size, and the pain was referred mainly to the instep. For five weeks the recumbent position was kept and antiphlogistics used. Her further history to the time I saw her, eight months after, was one of ups and downs, aches and pains. Suffice it to say that at my first visit I learnt that then a walk of a square would lay her up for a week with pain and weakness in the ankle and instep. She walked with a limp, and going up and downstairs was tedious and painful. When reclining, which was most of the time, the foot had a forlorn aspect, drooping forward and inward, and it admitted of too free passive motion in all directions except that of flexion, to which there was a yielding resistance, and this was felt by the patient as a disagreeable stretching sensation, not only in the calf muscles and those on the posterior aspect of the thigh, but also in the muscles of the back, even to the nape of the neck. Along the inner border of the foot pain was said to be constant, and at the articulation of the first metatarsal bone with the cuneiform, insertion of the tibialis anticus, there was too much mobility.

From six weeks after the injury to the time I was called she had had at various times three manipulators, non-medical people, who handled the limb as if they were afraid of touching it. There seemed to me to be no necessity for extreme gentleness in this case, as it was very evident the limb was now suffering more from impaired nutrition and innervation consequent upon the necessitated disuse than from anything else ; so I at once began to give it vigorous manipulation or deep rubbing as far as the knee, with brisk passive motion. That such could be done without causing any pain at the time seemed very wonderful to the patient, and no doubt the mental effect of this was good. I then asked the patient to move the foot herself as she was reclining, but it was a pitiful effort, scarcely visible. Nevertheless I said, "That's first rate. Now move the foot up and down and I will resist both ways." But this at first was a make-believe, for my so-called resisting motion was a simple aiding of the patient's voluntary

effort, and no resistance at all. The object was to encourage the
effort of the will, which in this condition might be one-half or two-
thirds what was necessary to do the required movement. Under
these procedures the defective will-power and impaired nutrition
improved *pari passu.*

After the first *séance* of twenty minutes it was to be expected
that such a hyperæsthetic individual would be quite tired out and
much worse generally, as was the case ; but I firmly adhered to
the theory that the more strength the limb gained the better able
would she be to bear her numerous aches, in which she agreed
with me. The patient was directed to try to flex and extend the
foot a few minutes twice daily when lying down. On the day
following the third massage, February 11, 1877, the patient walked
without limping any more. From this time onward, resistive
motion was no longer a make-believe, for it was evident the limb
was gaining strength from day to day, so that after a half a dozen
visits in eight days she went up and down a long flight of stairs
naturally and easily. At the end of ten days, after seven massages,
I tied a handkerchief around the ball of the foot (metatarso-
phalangeal joints), and put one loop of this over the hook of a
spring-balance, the indicator of which was pulled out several times
to 12 lbs. by the upward movement or flexion of the foot. The
other foot pulled 16 lbs. in the same manner; this one had also
been getting a portion of the same treatment, for it too had suffered
somewhat from disuse. In the next four weeks half a dozen more
massages were administered, and the foot and ankle gradually
increased in strength so that she could do anything with the limb
that her general strength or the state of her back would allow,
from going up-stairs with an elastic step or walking a half mile to
utter indisposition to move from the couch. The foot and ankle
were no longer the weakest parts which had to stand the strain,
the muscles of the leg were firmer and stronger, and it seemed as
if there was more of an arch to each foot, a result which an improved state of tonicity of all the muscles of the leg and foot, but
particularly of the tibialis anticus and peroneus tertius, might tend
to produce by supplementing any natural or acquired laxity of
ligaments. Six weeks later the patient wrote to me that "both

ankles were certainly stronger than before the accident last spring."

Similar results have recently been obtained from the use of massage in sprains and synovitis by Gerst, Wagner, Zabludowsky, Faye, Starke, Körner, Huillier, Fontaine, Witt, Estlander, Norström, and others. In five cases of acute serous synovitis, Johnsen obtained recovery by massage ; in forty-three cases of chronic synovitis, he obtained recovery in thirty-four and improvement in nine ; in eighty-nine cases of hyperplastic synovitis, fifty-five were cured, thirty improved, and four were without change. In fifteen cases of relaxation of joint capsules, recovery resulted from massage in fourteen and improvement in one ; in three cases of acute inflammation of the sheaths of the tendons recovery took place, and in six of a chronic nature, cure was also obtained. Sprains and contusions of the back from railway injuries or other causes (often erroneously supposed to be complicated with injury to the spinal cord) can be as appropriately treated by massage as when they occur to joints and muscles in other parts of the body. When acute symptoms have subsided this treatment should always precede attempts at motion. Dr. Herbert W. Page, however, advises motion under these circumstances, no matter how much it may hurt, evidently unaware of the effects of massage in preparing the way for this, and making it vastly more easy and agreeable.

CHAPTER XVI.

MASSAGE IN JOINT AFFECTIONS, CONTINUED.

COMPARED WITH HEAT AND COLD — STIFF JOINTS — HYDRARTHRUS — RE-
LAXATION OF MUSCLES — PERIARTHRITIS OF THE SHOULDER-JOINT —
BONE SETTING, IMPROPERLY SO-CALLED — DAL CIN.

"But Socrates, sitting up in bed, drew up his leg, and rubbed it with his hand, and as he rubbed it, said: 'What an unaccourtable thing, my friends, that seems to be which men call pleasure, and how wonderfully is it related towards that which appears to be its contrary, pain, in that they will not both be present to a man at the same time, yet if any one pursues and attains the one, he is almost always compelled to receive the other, as if they were both united together from one head. . . . Since I suffered pain in my leg before from the chain, but pleasure seems to have succeeded.' " — PHÆDO.

HEAT and cold have each been used with good results in recent and old joint-affections, sprains, and synovitis. Moderate heat causes a fluxion to the parts to which it is applied, dilates the vessels bringing the circulation, as well as those returning it. If long continued, it causes relaxation, and if of a high temperature, it acts like cold in causing contraction of the vessels and counter-acting vaso-motor palsy. Hence the plan of immersing a sprain in water at the temperature of 70° F. and gradually increasing it to the extreme point of toleration is excellent as far as it goes. But this is only a slight imitation of what massage does, for the intermittent momentary compression of the stroking and knead-ing causes a mechanical dilatation and contraction of the vessels (arterial, capillary, venous, and lymphatic) every time it is applied, from sixty times a minute and upwards, which is certainly sixty times oftener than could be caused by the variations of caloric in the same time; besides the aid to the returning circulation, by being pushed along by massage, is much greater than that caused by heat which tends rather to enlarge the area of stasis ; moreover, the pressure of the hand over effusions disperses them more rapidly than heat can do.

If cold be applied to a sprain or synovitis, this is all very well so far as the reduction of heat, pain, and swelling are concerned, but in place of the seat of injury being flushed and the returning currents hastened as by massage, the flow of blood is lessened and the outlet to effused products by veins and lymphatics is also rendered more impermeable in consequence of their contraction with all the other tissues that are cooled. Cold applications are not without danger, as they may convert inflammation into gangrene, and a lesser evil is that they may suspend nutritive action and hinder the process of repair to which moderate inflammation is necessary. Moeller has made known to us the result of Bauden's treatment of 500 sprains by cold water ; the average time for recovery was found to be 28½ days, over three times as long as that under massage,[1] thus :

104 cases recovered under cold water in from 8 to 20 days.
150 "   "   "   "   "   " " 20 " 30 "
110 "   "   "   "   "   " " 30 " 40 "
80 "   "   "   "   "   " " 40 " 50 "
56 "   .   .   .   .   .   .   not stated.

A few cases illustrative of the effects of massage in chronic local joint-trouble may now be considered in order.

In 1877, I attended Mrs. A. P., who eight years before had slipped and fallen, contusing and spraining her left knee. Synovitis set in, and the patient felt obliged to travel without rest or medical attention. For a year and a half the joint was painful and swollen, and at the end of this time it was severely flexed, and the patient was greatly run down. She then came under the care of a surgeon who built her up with rest and tonics, and later cut the ham-string tendons and used extension apparatus, so that, after a while, the patient could go about on crutches with

---

[1] Moeller, Du Massage, Journal de Médecine, Bruxelles, February, March, April, 1877.

a stiff bent knee. Her general health improved until she became quite well, though she was naturally rather delicate. On examination, I found the leg at an angle of 140° with the plane of the thigh, with but very slight motion in the direction of flexion and extension. On each side of the ligamentum patellæ, in front of the condyles, and also for several inches above the patella on the femur, there was a firm mass of hyperplastic connective-tissue which was greater and firmer in front of the internal condyle than elsewhere. There was no sign of superabundance of fluid in the cavity of the joint. Below the knee the cellular tissue of the leg was indurated and the capillary circulation languid. In October the patient had five massages, with special active and passive motion of bending and extending the knee a little farther than was comfortable, and the result was that the circulation became more active, and there was slightly increased freedom in the motion of the joint, with a·consciousness of greater strength. The patient was so encouraged that a month later she returned for a longer course of treatment. Massage, with tentative passive motion, was used every other day, and several times daily the patient sat with the well leg resting across the lower end of the femur of the stiff limb, while this was elevated as high as the chair on which she sat. At first, twenty minutes of this was all she could bear, but toleration increased, and at the end of a week the force of extension was increased by placing the leg on one of her crutches, the sole of the instep on the cross-piece for the hand, while the axillary support came under the knee, and the limb was strapped in this position, so that its own weight furnished power for extension by the length of a lever from the handpiece to the lower end of the crutch which rested upon the floor. This could only be tolerated for fifteen minutes to begin with; at the end of a week, for half an hour, and then the effect was increased by crossing the well limb over the other. The limb always manifested a good deal of trepidation from forced flexion or extension, the pain of which was referred to the inner and anterior aspect of the joint. Pressure on this spot with the thumb while forcibly extending the limb lessened the pain, and gave the patient a feeling of support and security. Hoping

to aid in rendering the hyperplastic tissue softer and more vascular, a flaxseed poultice was applied to it every night for a
week, but this was followed every morning by a painful lameness in the muscles on the posterior aspect of the thigh, and
even affected the glutei.   It passed away quickly under massage.
In three weeks we applied an ordinary curved ham-splint and
bandaged it on firmly.   The patient had never worn one of
these, and she was delighted, for in walking she could bear
more weight upon the knee than she had been able to do at
any time since it had been injured, and she could walk about
in-doors without the aid of crutches or anything else.   The
motion of walking with the splint made a uniform and vigorous extension of the limb, and this was very desirable, for the
tibia had a slight appearance as if it were subluxated backwards, but this was doubtless more apparent than real by reason
of the mass of tissue on the anterior aspect of the joint.   In
five weeks we had gained one and one-eighth inches extension
at the ankle, as shown by an outline, and the thickened tissue
had decreased, showing one inch less of circumference above
the knee and three-quarters of an inch less below and in front
of the condyles, the former circumference being the same as
the well knee, the latter still seven-eighths of an inch greater
than the same circumference on the other knee.   The ham-splint
continued to afford great security and comfort in walking.
Three weeks later, the campaign against the knee was practically at an end, on account of the appearance of cough, headache, backache, and pains in the lower part of the abdomen.
General massage, rest in bed, light nutritious food, and tonics
soon restored the patient to her customary health, and then she
departed for a more genial climate further south.   Nothing
more was done for the knee, and she has not yet got rid of her
crutches.   But there is no loss without some compensation, for
the use of crutches in this case, in all probability, strengthened
and expanded a somewhat delicate chest, as well as imparted tone
to the muscles of the abdomen, thus supplementing the work of
the uterine ligaments, for there was a history of their relaxation,
and enlargement of the uterus.   With a stronger constitution,

massage, movements, and mechanical appliances would have been·
more successful. We may infer this not only from the history of
the case, but also from the result in the following case :

A man, thirty-three years of age, after a contusion of the right
knee in 1870, had stiffness and enlargement of the same remaining. On the 15th of August, 1874, he cut himself with an axe
in the same knee, causing a penetrating wound of the joint
which closed in three weeks ; but anchylosis began to form, which
without doubt was still membranous, as slight mobility was
present. The circumference of the knee over the head of the
fibula was 32 centimeters, at the lower margin of the patella 38,
at the upper 40, and a little above at the end of the cicatrix 37
(against 30.5, 35, 35, 34 centimeters of the well knee). The cartilages and apophyses may also have been involved in the hyperplastic process. After massage for fifteen minutes, the circumference of the knee had decreased one centimeter. After five
sittings, the enlargement had decreased considerably, and motion
had increased, but the treatment was then interrupted on account of the appearance of furuncles. When these had healed,
the condition of the limb in general had considerably improved,
but the swelling was much the same. After each massage, motion became freer and swelling less, but after a few hours the
swelling and stiffness returned. Yet there was gradual improvement, especially when the patient took half an hour's active exercise after the massage. Once more the appearance of
furuncles interrupted the treatment. After three months, forcible flexion under chloroform was done, and the breaking of adhesions could be distinctly heard. After this flexion was natural. The joint was abundantly *masséed* immediately after the
operation. The following day the patient could walk around
without hindrance, and after about a week, during which he
had daily massage, the circumference of the knee had decreased,
but was still larger than the other, and the patient could walk a
long distance without limping or getting fatigued. Several
months afterwards, there was no relapse.[1] Milder cases than

---

[1] Westerlund in Finska läkaresällsk. handl., xvii., 3, Oct. 4th, S. 144, 1875.

these of chronic hyperplastic joint-trouble yield more readily to massage.

Massage may prove valuable in chronic effusions into the joints, whether painless or painful, whether dependent upon increased secretion or lessened power of absorption. In these cases there is generally thickening of the capsule. Barwell calls attention to the fact, and Billroth expresses himself in like manner, that frequent application of blisters and stimulating embrocations often relax and injure the skin, producing therein a state of chronic congestion, a passive hyperæmia and thickening, similar to the diseased condition they are intended to combat, but which they frequently aggravate. Issues and moxas may inflict similar injury. After acute synovitis and when lingering inflammation has been subdued, Barwell says massage and passive motion should be resorted to in order to promote absorption of new growths. He regards it as very valuable for superficial joints, as it often restores flexibility and perfect shape to the joint more rapidly than any other means with which he is acquainted.

Mr. L. A., twenty-seven years of age, and well built, had suffered from slight effusion into the right knee-joint for five years, which was suddenly converted into a large effusion by a contusion of the knee from a fall while skating, eight weeks before he came to me. The effusion had been reduced considerably by means of compression, but still bulged out a good deal on each side of the ligamentum patellæ, and on each side of the tendon of the rectus femoris. Motion was natural in all directions except extreme flexion. The knee felt weak and stiff. There was no evidence of acute inflammation, and no pain. The case was one of painless effusion within the capsule. In addition to the compression, I gave the patient seven massages in three weeks, and the knee steadily improved, the effusion decreased, and the muscles gained in size and firmness. At first, there was pain only on pressure at the anterior aspect of both condyles, but this disappeared. After two massages, the patient resumed full work all day, engraving, and in the evening walked one mile and a quarter on an icy sidewalk without detriment. Previous to this he was doing but half a day's work, and walking only a few squares at a time. For five months he disappeared and then returned to me much dispirited, with the knee worse than when I last saw him. The effusion was nearly all gone, it is true, but the limb could not be fully extended as before, and there was more stiffness and occasionally crepitus with spontaneous pain. Counter-irritants had been freely used in the meantime, and all the tissues from the skin down were hard, stiff, and dry like sole-leather; just in the opposite condition from that which massage induces, namely, suppleness and elasticity, while stimulating absorption to a high degree.

Massage has been found to work well in recent cases of hydrarthus, "water on the knee," but with chronic ones like that just narrated, the absorbent vessels are no longer in a normal condition, but are more or less obstructed or over-burdened. It might be prudent to try massage with compression for a while, and if nothing were accomplished, aspiration might be resorted to and then followed by the means that preceded it. Prof. J. Nicolaysen (in *Norsk. Mag. f. Lägevidensk.* 3, R. IV., 3, 8, S. 124, 125, 1874) communicates the following cases: A man, thirty-two years old, had suffered from hydrarthus for six and one-half years. Repeated puncture and evacuation had always been followed by a reaccumulation of the fluid. Massage was used for several months and the patient returned to his work. There was no relapse as formerly. A man, fifty years of age, had suffered from hydrarthus of the knee-joint for four months. After the use of massage for seventeen days the collection of fluid in the knee-joint disappeared, but the swelling of the capsule and sensibility of the external condyle continued for a long time. After six weeks' treatment the patient was discharged cured. In another case where the affection had lasted for ten years, improvement was obtained by means of massage, but the patient left the hospital before treatment was ended, with the capsule thickened and the surrounding tissues relaxed. Norström and Professor Gussenbauer have had similar encouraging experience with the use of massage in *hydrops articulorum.* After narrating an exemplary case, Gussenbauer concludes by saying: "I could still tell you, gentlemen, of several more cases of *hydrops genu*, in which by means of compression with sponges and massage this obstinate affection has been perfectly removed, or so much improved that the patients could resume their occupations." In fourteen cases of *hydrops genu*, Zabludowsky obtained good results from massage; in recent cases, in from six to eight days; in old ones, in from one week to five months.

Only from such a careful and distinguished author as Barwell on "Diseases of Joints," is an opinion like the following of value (pp. 150, 151): In his chapter on strumous synovitis he says, "We may only be called upon after the patient has suffered for some considerable time; has been kept in bed with perhaps an issue that has been open for six weeks or two months, or possibly with no treatment at all. The joint will probably be found shapeless, swollen, pulpy, perhaps it may be painful; probably, particularly if the knee be in question, it will be a good deal flexed. Now, we shall in nearly all cases find on examination, unless the disease has gone too far, that the whole joint may be manipulated without producing pain; that pressure upon the choice seat of tenderness will cause no expression of suffering, and that no startings or any acute pains disturb the patient's sleep. Even in such a case as this, we may in all likelihood cure the patient by first applying strong pressure, manipulations, rubbing, and passive motion. The condition into which the new tissue has fallen is simply a passive one; the material exists, but there is no action in it; perhaps there may have been an abscess which has left a sinus, but the suppuration is very sluggish; the rest of the tissue is doing nothing.

"Now, if the granulations be allowed to remain in this passivity, they may, after some years, contract and consolidate even *in spite of such treatment;* but their more general course is to take on a retrograde action, gradually to yield to suppuration, and to involve the textures of the joint which they inclose. Our object should be, taking advantage of the passive state, to produce absorption of the jelly-like tissue. The painless condition upon pressure, and particularly of that spot which is the chosen seat of tenderness, is the proof that we may employ not merely pressure and massage, but passive motion ; and we can, in a great number of instances, even after abscesses have formed, produce absorption of a large portion of the false tissue and consolidation of the rest. I desire to lay powerful stress upon this point of enforcing passive motion as soon as actual inflammation is checked, and mere vegetative cell-growth is the only action going on. M. Bonnet, the first writer who attempted to show the value of such means, has not limited its use sufficiently to the cases of which we are now treating. The counter-indications to this treatment are an active condition of the swelling, evidenced by pain and tenderness, any considerable amount of degeneration or suppuration, starting pains and tenderness of the joint surfaces." Billroth also recommends massage in torpid cases of *tumor albus.*[1]

The following case, I think, clearly shows the benefits of massage and special exercise in relaxation of the quadriceps extensor mus- ·cles resulting in part from a former contusion and synovitis. In 1880, Miss M. B., then thirty-seven years of age, came to me. She was of large frame and well nourished, but her muscles were flabby and her movements not easy and graceful, owing to some extent to the length of her levers. At fourteen years of age, she fell upon the ice and hurt her left knee, and ever since she has had some trouble with that limb. The history and symptoms showed relaxation of the quadriceps extensor muscles, allowing the patella to glide externally without the voluntary or involuntary power of immediate replacement, and this would cause the patient to fall, and then she pushed the patella into its place with her hand. The knee was sensitive and easily jarred, but otherwise normal. For relief and prevention of displacement of the knee-pan, she wore an elastic envelope with a hole in it into which the patella fitted. Tendon reflex was normal. When she was lying down, she could not raise the limb off the couch with the knee extended, but when some one else elevated it, she could then hold it so for a few seconds without support. After two sittings of massage, percussion, and assistive movements, the pa-

---

[1] Surgical Pathology, p. 750.

tient could raise the leg extended without assistance, and after five visits she could do this easily. Every now and then the patient had massage frequently for a few weeks. Two months after her first visits to me, she was in the habit of leaving off her elastic support and not only walking about without it, but also going up and down stairs in the dark, in the duties of her profession, which was that of a nurse. Here is the record of her visits to me at the end of a year, from the first time I saw her. November 25th, 1881. After twenty minutes of massage, patient elevated the limb extended at knee and held it so for one and one-half minutes. After ten minutes more of massage, she held the limb out straight for two minutes. November 26th. After twenty minutes of massage, limb held up for two minutes; after ten minutes more of massage, for three minutes ten seconds, without much fatigue. November 28th. After twenty minutes of massage limb extended and elevated for three minutes ten seconds, and after ten minutes more, for three minutes twenty seconds. Massage on November 30th, December 1st, 3d, and 6th,. when after twenty minutes of massage, limb was held out and up for four minutes. December 9th. Limb extended and elevated for four minutes and five seconds; December 12th, seven minutes. The use of the leg had improved proportionately in every other respect. I would not consent to this patient going without a few turns of a roller bandage for protection against slipping of the patella. I have an impression that she wrote me a few weeks later that she could hold the leg out straight for ten minutes at a time, for I had advised the continuance of this exercise from the first.[1] If any one will try to hold his whole leg and thigh thus extended horizontally without support for one minute by his watch, he will appreciate what was accomplished in this case. Previous to her coming to me, she had been advised to lie still with the leg on a ham-splint for six months, so as to give the quadriceps extensor a chance to grow shorter. By means of massage, her occupation was in no way interfered with, and the natural contractility of the muscles was greatly increased. Relaxation of

---

[1] Since writing the above I have received another letter from this patient assuring me that my impression was correct.

ligaments is often diagnosticated when the condition is really one
of atrophy and relaxation of muscles allowing undue traction upon
the ligaments of the joint. If space permitted I could give further
illustrations of this.

Periarthritis of the shoulder-joint, a subacute or chronic inflammation of the subacromial bursa and of the loose areolar tissue under the deltoid, with thickening and the formation of ad-
hesious entangling nerves and tendons, hindering motion and
setting up neuritis while the articular surfaces are in a normal
condition, is a very stubborn affection. Surgeons in extensive
practice have expressed the opinion to me that it seems to be
on the increase of late years. This would seem to indicate an
increasing constitutional predisposition to the affection over and
above the usual immediate causes : injury, rheumatism, catching cold, or prolonged immobility. Those of traumatic origin
do the best under treatment. The main impediment to motion
would seem to be the thickening of the walls of the subacromial bursa, which prevents the gliding of the superior extremity
of the humerus under the acromion. Besides the muscles being
atrophied from disuse, one fact, I think, has been overlooked,
that they are frequently in the state known as myositis — tender, sore, and indurated. It is generally agreed that the let-
alone treatment of these cases allows them to get worse, and
favors anchylosis. The most rational plan of treatment consists in the use of massage, passive motion, electricity, and
douching. These may be used all at the same time, or one or
the other will be more insisted upon, according to the nature of
the case. Massage and passive motion prevent the formation
of adhesions, and loosen to some extent those that have already
formed ; therefore in the early stage of this affection it may be
both preventive and suffice to bring about recovery. After
firm, deep adhesions have been broken up under anæsthesia,
their reformation will be in part or wholly prevented by means
of massage, and the immediate soreness of the tissues around the
joint diminished, as in the case of a sprain ; for this is really
what the healthy tissues have to suffer while the adhesions are
being ruptured. This will be more apparent when we call to

mind that growth has been going on in these healthy tissues
with limited motion, so that some of them are shorter than natural. The full extent of motion and exercise compatible with
firm adhesions can only be ascertained and cultivated by means
of massage and passive motion, followed by active motion. On
this account, a course of this sort, preliminary to the operation
of breaking up adhesions under anæsthesia, would seem to be
commendable, so that muscular fibres that are glued together
by lymph might be set free, and so much relaxation gained.
While using passive motion without anæsthesia, I have usually
found that the muscles can be better relaxed and more strain
put upon them and the adhesions, and these gradually stretched,
by proceeding gently and tentatively, than by sudden and brisk
jerking and pumping, which only makes the muscles contract
all the more stubbornly in order to protect the parts from pain.
In general, the result of the treatment in these cases is far from
satisfactory. The mobility of the scapula makes up in great
part for the lack of success in restoring motion to the scapulo-
humeral articulation, and this compensatory mobility can be increased by massage and movements.

For warding off the impending stiffness and relieving the pain
and swelling that appears in the first stage of periarthritis of the
shoulder, I believe the value of massage can hardly be overestimated, for in doing this it prevents subsequent mischief that
cannot easily be overcome. A physician next door to me fell
and injured his shoulder-joint. He called on me the following
day. I found great periarticular swelling, ecchymosis and pain.
At the end of a week, after daily massage, the joint was well,
and has continued so ever since, now eight years. Miss M.,
forty-two years of age, weighing one hundred and fifty-four
pounds, fell backwards and sideways, striking the posterior aspect of the right shoulder-joint just over the circumflex nerve.
The shoulder swelled so that she could not get her dress on,
but this decreased to its natural size in about five days. It gave
her pain at night, but scarcely any during the day. She came
to me two weeks after the accident. The joint was of natural
appearance, allowed free passive motion, but there was paresis

of the deltoid, so that she could only move the arm voluntarily a few inches from the side. I gave the shoulder massage and faradization, and after the first visit it was more comfortable, and allowed her to sleep better. On the third day, at the third visit, after faradization for five minutes, she could elevate the whole arm, and circumduct it naturally, but feebly, and the strength of doing this was increased by massage and percussion, but the increase of power was only retained for about twenty minutes. It gradually increased at and after every visit at which massage, percussion, and electricity were used, until at the end of nineteen days, after fifteen visits, the patient could put up a six-pound dumb-bell, and a week later she had quite recovered. In two weeks after this patient came to me, she was free from pain at night, and could elevate the arm horizontally before manipulation and electricity. Anxious to test the comparative effects of massage, percussion, and electricity in this case in restoring the power of the deltoid, I repeatedly varied them at different sittings, using one first at one time, another at another, and always got the same results. Five minutes of faradization had a greater effect than five minutes of massage, and one minute of percussion with the ulnar border of the fingers separated had a greater effect than either. The effect of each was increased by being followed by the others. False crepitation [1] in this joint was very marked before a sitting, but much less afterwards. The patient was well for two weeks, and would have continued so, but while heated and perspiring, she was exposed to a draught of cold air, which brought pain and stiffness in the shoulder. For this she put on a belladonna plaster, which, besides producing symptoms of belladonna poisoning, caused great swelling, irritation, and œdema of the whole arm and hand. When these had passed away, it was found that the shoulder-joint had become worse, so much, indeed, that an expert advised motion under ether. This was not consented to, so we used massage, electricity, and passive motion industriously every day for two weeks, repeating the same interesting ex-

---

[1] This was observed in two cases of every ten mentioned by Duplay. Archives Gén. de. Méd., Nov., 1872.

periments as before, with the result that in two weeks almost perfect motion was gained, and the joint was then equal to all demands upon it, and no more complaint has been heard from it.

Mr. C. A. had the structures on the anterior aspect of the left shoulder-joint severely stretched by the sail of his yacht changing position, so that the whole weight of his body came suddenly upon them, while the arm was jerked upwards and backwards. Prof. David W. Cheever sent him to me three months after the injury. The patient stated that pain and limitation of motion had been increasing ever since the accident. Under massage and vigorous passive motion, he made a good recovery.

Howard Marsh, of Saint Bartholomew's Hospital, reports the case of a gentleman who fell from his horse, bruising his arm and dislocating his shoulder. Two months afterwards he came to London with the joint almost stiff. (It had been set immediately after the accident.) The adhesions were broken up, but at the end of a week the joint was as bad as before. The operation was repeated with no better result. He was then told that he would have a stiff shoulder for life. This he did not accept, but consulted another surgeon who again broke up the adhesions, and consigned him immediately to a manipulator, who *masséed* the joint for six weeks, and at the end of this time the stiffness had all disappeared, and never again returned. Dr. James J. Putnam, in commenting on this case, wonders whether what was called the after-treatment ought not to be the primary treatment, and whether it might not have been effective if applied as thoroughly in the first place.

With regard to the effects of breaking up adhesions under ether in these cases, Dr. Putnam thus gives his experience of twelve cases: "Those operated on once and cured or benefited, six in number. Those operated on and not cured, within the time they were under observation, five in number. Those operated on once or more without benefit but eventually cured by other means, one in number" (*Boston Medical and Surgical Journal*, Nov. 30th, 1882).

Norström reports three cases of periarthritis of the shoulder-joint that were benefited sufficiently by massage to give them full use of the affected arm.

Where motion is very limited in a joint and cannot be increased, massage may be continued indefinitely with benefit as a substitute for exercise.

Cases of joint trouble that give the least response to massage are those where no objective points can be found, though I have had good results from this treatment in neuralgia of the joints, after all other methods had come to an end with little or no benefit. Such patients will quickly place more confidence in the physician and his abilities than he does himself, if they can only see that he understands their case, and he should make the most of his influence in encouraging gradually increased exercise, a most valuable preliminary and concomitant of which is massage. Rubbing by an automaton in such cases is useless, — intelligence and will must go with it.

A patient with an old joint-malady often goes to a *bone-setter*, as he is popularly called. No matter whether the joint be stiff or lax, enlarged or of normal size, the diagnosis is at once made that a bone is out. The patient looks sceptical, but the bone-setter does not care whether the patient believes him or not, he is confident that he can cure the joint all the same, and so assures his patient. Allowed to proceed, a sudden movement or two accompanied by a snap convinces the patient that the bone-setter was right, who tells him that was the noise of the bone going back to its place. He is told to use the limb and leave off his supports. This is attempted, and if successful, then it appears as if a miracle had been performed, and the patient lets everybody know of it; but if not successful, or if the patient has been made worse, which is often the case, then be sure said patient keeps very silent on the subject.

A sudden movement of any joint, healthy or diseased, will usually cause it to snap; and if there be old adhesions, this is an excellent procedure. But the noise of adhesions giving way is different from that caused by a bone slipping in or out of place. Dr. Wharton P. Hood has described the method of bone-setters in his book, "On Bone-Setting, So-Called." But he makes these procedures applicable to too great a variety of cases. Hood states that the cases which bone-setters benefit by breaking adhesions are those of joints in which there is a slight degree of mobility checked by pain, a spot tender on pressure, and an absence of acute disease. These symptoms are also found in joints in which there are no adhesions; for an increased involuntary tension of the muscles occurs in joints that are injured or diseased, which the force of habit often causes to continue after the joint is well. This involuntary tension keeps the joint in an irritable condition and limits passive motion, hence the above symptoms. In some cases it is sufficient to explain to them that they may relax their muscles by repeated voluntary effort; in others massage with gentle, persuasive, passive motion will greatly aid recovery. Later, to strengthen the muscles and to teach the patient how to use them, gradually increasing resistive motion, is of value. It is a mistake to suppose that violent rubbing is of use in such cases; it begets reflex contraction of the muscles, increasing the evil it is intended to remedy, and causes an intense hyperæmia and hyperæsthesia of the skin besides chafing it. To avoid the last objection and to conceal the manipulator's ignorance, oily substances are made use of.

I have used vigorous passive motion in several cases, notably in three cases of stiff ankles resulting from fractures of the lower end of the tibia extending into the ankle joint. The adhesions were heard and felt breaking, the pain was momentary, and two of the cases could walk immediately after without crutches, the freedom of motion in the joint being doubled. The third walked lame without any support before I saw the case. Though some adhesions were broken, but little increase of motion was gained; the improvement in walking and in going up and down stairs was much greater than one would have thought possible from the slight increase of motion in the joint. In this last case adhesions within and without the sheaths of the tendons may have been loosened by the accompanying use of massage. In such cases excellent preliminary measures to the breaking of the adhesions are a warm bath followed by the soothing influence of gentle stroking of the joint and limb, and by deep manipulation (without slipping of the fingers), which has a decidedly anæsthetic effect. Gentle tentative passive motion may then be tried so as to judge of the state of the joint and the amount of force to be employed. After the violent passive motion, deep kneading and a tolerably tight bandage increase the patient's comfort. When necessary, repetitions of these efforts are better tolerated than the first in suitable cases.

Hood to the contrary, violent passive motion would not be appropriate for recent sprains, and seldom for rheumatic and gouty joints.

It is but a few years since the Italian peasant woman, Regina Dal Cin, visited this country, and by reducing dislocations that never existed made a great reputation among those who knew nothing about such matters. One physician went so far as to give her credit for doing massage well, but neither he nor she have ever given any proof that they knew anything about massage. A patient of mine, well advanced in years, slipped off the curbstone, and came down "*kerchunk*," stretching the arch of her foot beyond its natural limits. Massage, tight bandaging, and a stiff sole had been of great service to her, so that she could finally walk about, but not so easily as before her mishap. Fond of indulging in *advanced notions*, and pitying the bigotry of the regular physicians, she sent for Dal Cin, who, supposing the patient could not walk at all, at once set to work to make her do so by pulling away at the great toe, and pushing at the inner part of the sole. After a little while spent in this manner, she asked her interpreter to tell the patient that the ligament leading from the great toe to the heel (!) had been out of place, but she had put it back again so that the patient would be able to walk. The patient got up and walked forthwith to the joy and astonishment of Dal Cin and her interpreter. Observing their emotion, the patient coolly told them she could walk just as well before Dal Cin came.

Another patient recovering from hip disease, whose limbs were of equal length, went to Dal Cin and had the joint set. It had never been out of place. The reputation she gained among "*advanced thinkers*" by this case was astounding. The boy repeatedly showed me that he could walk without his cane before he went to her. The report was that he could not.

In a letter from Vienna, published in the *Boston Medical and Surgical Journal* of February 1st, 1872, by Dr. James R. Chadwick, an interesting account is given of Dal Cin, which is as follows : " For a year or more she has been moving about from village to village, and town to town of Northern Italy, Hungary, and Austria, claiming to effect, by certain manipulations, a perfect cure in all cases of dislocation of the hip-joint, and such like diseases. The number of those treated by her with perfect success is said to have been many thousands. At last, two months ago, she ventured into Vienna, and began to ply her trade, and soon acquired such renown that a commission [1] from the medical faculty was appointed to investigate her claims to be admitted to practice in that city — a privilege granted only to those who had proved themselves fully qualified. The four cases seen by the committee, in all of which the head of the femur could be unmistakably felt in the acetabulum, were all pronounced by Dal Cin to be dislocations which she would proceed to reduce. To effect this, she first placed the patient so as to show the greatest possible deformity,[2] then seized the limb and made several painless and apparently complicated motions without fixing the pelvis, and afterwards by means of the shortened limb, drew down that side of the pelvis, laid the feet together, and held them so, while she showed to the astonished public that the soles were side by side, and that the limbs appeared to be of equal length.[3] In order not to be detected while preparing the bandage, she bent the knee of the long leg, and placed the foot across the back of the other. A bandage of tow and white of eggs was applied to the thigh with an external splint. The patient was then ordered to remain in bed, and not remove the dressing for a month. The reduction being announced as perfect, the physicians proceeded to measure the

---

[1] Having the curiosity to see whom this committee was composed of, I looked in the Allgem. Med. Woch., and in No. 47, 1871, I found them to be Professor Weinlechner, Dr. Nusser, city physician, and two physicians-in-chief from one of the public hospitals, Drs. Lorinser and Mosetig.—D. G.

[2] Any well person may be so placed.—D. G.

[3] A patient can easily be placed so that a short leg will appear as long as the other.—D. G.

limbs, with the result that in none of the cases was the slightest change in length, position, or mobility discovered. After these experiences and proved facts, the commission felt themselves justified in publishing the following statements:

"'1. The Frau Dal Cin has not the most superficial conception of a dislocation, or the means of its reduction. In one case she called the great trochanter the head of the femur.

"'2. The attempted operation consists in irregular, changeable, painless passive motion, insufficient to reduce a dislocation or stretch a contraction.

"'3. The success of these painless motions was in every case *nil*.

"'4. Inasmuch as Dal Cin, after the so-called reduction, makes use of certain manœuvres with the object of deceiving as to the comparative length of the limbs, it follows that she is fully cognizant of the deception, and that she is, in the true sense of the word, a female swindler.'

"As a result of this report she has been forbidden the city, and driven to practice among a more credulous folk." The people of Vienna were certainly credulous enough before the proceedings were investigated.

Mr. Howard Marsh, who is greatly in favor of forcible movements for loosening adhesions in suitable cases, shows the dangers incurred by employing bone-setters.

A patient sixty years of age, with malignant disease of the pelvis which had caused plugging of the iliac veins and pressure on the sacral plexus, was told that his hip was out. The joint was said to be "put in" by wrenching, a form of treatment which greatly increased his suffering.

A patient with a large sarcoma of the muscles of the thigh just above the knee, was told that the knee was out and must be put in. Arrangements were made for the operation, but fortunately other advice was taken, and the proceeding was declined.

In a case of far-advanced angular curvature of the dorsal spine in a little girl, the "buttons of the back" were said to be out. The treatment adopted terminated in the child's death.[1]

---

[1] St. Bartholomew's Hosp. Reports, vol. xiv., 1878.

The less people know the greater is their faith, and hence the greater is their enthusiasm, which is so necessary to convince and carry along others. The most learned people outside of medicine are the most easily gulled, because they presume to know what they do not know, and one assertion is as good as another with them. Child-like simplicity, good-natured enthusiasm, and unbounded sympathy, are the three prerequisites of a successful *magnetizer* or *bone-setter*, and the mysteriousness of their doings making a deep impression on their patrons, all other attributes are added unto them. What is testimony worth under such circumstances ?

Reference to the surgical uses of massage would not be complete without mentioning that in incomplete or soft union of fractures, when a fixed dressing has not brought about consolidation in due time, it is now considered good practice to resort to friction of the fractured ends against each other, either by allowing the patient to walk a little or by deliberate manipulation in the hands of the surgeon. This induces a condition similar to that which occurs at first in an ordinary fracture, and success will be more likely to accompany the renewal of a fixed dressing. But when this and other methods fail, such as sub-cutaneous puncture, injection of irritating fluids, wiring, &c., or are objectionable on account of the fracture being in the vicinity of a joint, percussion over the fracture has been employed with brilliant results. This method has proved to be both safe and effectual. It is done with a metallic mallet faced with India-rubber, for five or ten minutes at a time, once in forty-eight hours or so, until pain, heat and swelling show that active hyperæmia and a renewal of the reparative process have set in. It is not applicable to all cases, but to those of non-union, where the ends of the bones are in apposition and no interval of deficiency nor any tissue intervenes between them. It has been used successfully in cases of fibrous union when absorption and attenuation of the fragments or eburnatina had not occurred, and when there was no constitutional dyscrasia. Dr. H. O. Thomas, of Liverpool, is said to have been the first to employ this method.[1]

---

[1] Braithwait's Retrospect, 1876, New York Med. Journal, February 6th, 1886.

In intra- and para-articular fractures M. M. Lucas, Championnière, Tripier and Rafin have found that immobilization is accompanied with certain dangers, whereas massage acts well from the first and can be used with other means. The dangers of immobilization are stiffness of the joints, atrophy of the muscles, and the ischæmia of the muscles caused by the pressure of fixed dressings may be mistaken for neuritis. The advantages of massage are that it promotes absorption of effused products, prevents stiffness of the joints and atrophy of the muscles and favors repair. It does not delay the period of consolidation, and where this is possible it has hastened union in many cases. Massage should be used once or twice daily. M. Championnière found that its employment relieved pain as most observers do, but M. Rafin found that its application was painful. The manner of doing it would probably account for the difference. Good recoveries were obtained by each. Cases in which there was no displacement did well under massage and passive motion without immobilization. Those in which there was displacement of the fracture, had massage for a few days at first to hasten absorption, then a retentive dressing for the briefest period possible to make sure that the displacement would not return, and after this massage and passive motion for the restoration of mobility. As soon as consolidation had taken place there was free and easy motion, a concurrence not sufficiently emphasized by our observers in their otherwise excellent remarks, to which it is impossible to do justice here. Of the cases recorded by M. Rafin,[1] recovery took place in three of fracture of the fibula in 13, 22 and 13 days respectively; in two of the radius, in 19 and 20 days respectively; in a double fracture of the ulna, in 27 days; in a fracture of the external condyle of the humerus, extending into the joint in a child, in nine days; in one case of fracture of both malleoli in a child, in 15 days; in one case of fracture of both malleoli with subluxation of the foot outwards and backwards, in 40 days. Few men would have had the hardihood or the patience to use massage in a case like the last until consolidation had occurred.

---

[1] *Lyon Médical*, March 18, April 1, 8, 15, 1888.

Even when good union has been obtained of transverse fracture of the patella by means of immobilization, or by suturing the fragments, or by fixation of them with Malgaigne's hooks, the resulting stiffness of the knee-joint, atrophy of the quadriceps extensor and impaired motion may cast a shadow over the most careful treatment or the most skillful operation. But the lesson taught by cases in which the fragments have remained widely separated and yet accompanied by excellent motion has not passed unheeded; for Prof. Tilanus of Amsterdam treats cases of fracture of the patella without immobilization, using instead compression, massage, and early movements of the joint, leaving consolidation to take care of itself. By this means the effusion is quickly dispelled, atrophy and stiffness prevented. The patients are encouraged to walk after the first week. In six cases that Prof. Tilanus treated this way the patients could walk very well in fourteen days. In one case that M. Rafin treated there was excellent motion in 42 days, the patient walking perfectly and experiencing slight difficulty only in descending stairs. The seat of the fracture could then be felt with difficulty, being marked by a slight depression of the skin, and the leg could be flexed within two fingers' width of the thigh.[1]

Dr. Wagner, a regimental surgeon in the Austrian Army, obtained astonishingly good results in five cases of fractured patella, which he treated with massage, active and passive movements, but without even the application of a bandage. His method is as follows : The patient is put to bed and the affected limb stretched on a simple inclined plane or an adjustable wooden splint in such a way that the heel is elevated. For the first three or four days an ice-bag is placed on the swollen and painful knee. As early as the fourth day, massage may be begun not only over the entire extent of the quadriceps muscle, but also over a large portion of the knee-joint. This is done daily and soon followed by passive movements. As soon as possible without causing great pain, the patient himself begins to move the joint, and after the lapse of 14 or 20 days may attempt to walk on crutches, and in six weeks is usually able to walk without support. Should the fragments be

---

[1] *Vide* the very instructive article of M. Rafin in the *Lyon Médical*, Sept. 5, 1886.

widely separated, Wagner recommends subcutaneous suture ; but even here massage and gymnastics cannot be begun too early.[1]

It is worthy of note that the principal muscles of the quadriceps extensor that effect extension of the leg are the vastus internus and vastus externus. The insertion of these on the lateral aspects of the patella is much lower than is generally supposed, and affords leverage to extend the leg when an upper fragment of the patella is detached. Rupture of the rectus femoris alone has little or no influence in hindering walking.

When the knee-joint is healthy even total extirpation of the patella impairs its motion but very slightly, and that only in a relatively weakened power of extension of the leg. Dr. Kummer of Geneva has reported several cases illustrative of this.

After union of any solution of continuity, whether of bone, muscle or nerve, massage and movements are in order ; and even while repair is going on the mobility of the joints may be made still more secure by altering the position of them at each dressing, as of the fingers in fracture of the forearm, or of the forearm in fracture of the elbow.

M. Marevéry in an article in *L' Union Médicale*, February, 1889, on fracture of the fibula treated by massage and mobilization, says that almost all of his patients treated in this way received the use of their limbs in less than 19 days after the injury. This treatment can be used only for certain kinds of fractures, such as those of the inferior extremity of the radius and of the inferior extremity of the fibula. Cold applications and a rubber bandage were also employed, as well as the massage from the first, and passive motion was given to the joints early.

---

[1] Wagner, V., in *Deutsche Med. Zeitung*, Feb. 6, 1888.

CHAPTER XVII.

" I'll knead him ; I'll make him supple." — SHAKESPEARE.

RHEUMATIC gout, rheumatoid arthritis, or *arthritis deformans*, is too familiar to us all to need any lengthened description. We readily distinguish it in most cases by its slow and progressive character, by its chronic or subacute course with frequent exacerbations and remissions ; by its spreading symmetrically from the small joints to the larger, usually in the upper extremities first ; by the thickening, permanent enlargement, and often deformity it occasions ; by its never affecting the heart, and not being hereditary ; and by its attacking females oftener than males. The term rheumatism is too often used to cover a multitude of diseases, as charity is to cover a multitude of sins ; and when rheumatism assumes a chronic form after an acute attack, it may be impossible to see any difference between this and rheumatic gout. Recent views regard arthritis deformans and kindred disorders as possibly of nervous origin. This might seem to be the case when the malady begins with a quiet pulse and clean tongue, and when the patient complains of languor, loss of appetite, pain, stiffness of the joints, and perversion of nutrition. But as to causation, the ground may be easily covered by saying that whatever deteriorates the system, whether by mind or body, in those predisposed to this or any other disease may, and probably will, precipitate its onset.

It is to be regretted that our success in the treatment of this obstinate affection is not equal to our knowledge of its pathology. An ounce of relief or an extra inch of motion is worth infinitely more to a patient than a foolscap of information to the effect that the disease is a panarthritis involving cartilage, bone and synovial membrane, ligaments, tendons and bursæ, with thickening of the articular lamellæ and thinning and alteration of the cartilages. With strange inconsistence Prof. Senator, of Berlin, in his Treatise on Diseases of the Locomotive Organs, describes this malady as usually selecting joints which are most continuously and severely overtasked, as in sewing, knitting, watchmaking and the like; while on another page he tells us that the thumb is generally spared and remains freely movable, as if it were not tasked as much as the fore and middle fingers. For such a description he may be credited with drawing upon his imagination rather than on observation. I am sure we all know that the freedom of motion is longest preserved in those fingers which are used the most: first, in the index; second, in the middle finger, while the thumb shows greater signs of being affected, and the ring and little fingers are generally worst of all. Moreover, Senator's description is again at variance with the only remedial course of which he speaks in unequivocal terms by saying: " It is important in all the forms of this disease to maintain the functional activity of the affected joints, as far as possible, by means of active and passive movements. Absolute rest promotes stiffness of the joints, fixes the limbs and atrophies the muscles." If his first statement were correct, then the treatment he so strongly recommends would be veritable homœopathy; disease produced by motion, relieved by motion. But unfortunately this is not so, at least with regard to dosage; for in using massage and passive motion in these cases, long and frequent visits and arduous work are necessary; but in my experience amply repay both patient and physician for time and trouble expended. Led on by gradual improvement from the use of massage in five out of six cases of well-marked rheumatic gout, I kept up this treatment until unlooked-for results were obtained, so that four of these cases regained tolerable use of the affected limbs, and in one recovery seemed to take place. Berghman and Helleday report three cases, Cronfeld one case, and Balfour, of Edinburgh, two cases of rheumatic gout treated by massage and movements, with results similar to those I have obtained, and only one of these had a disappointing relapse. The mode of procedure in my cases was deep manipulation without friction or inunction, passive motion as far as pain would allow and sometimes farther, and resistive motion as soon as it could be done, the details of which I have described more fully elsewhere. If pain disappears soon after it is caused by any of these operations, it may be disregarded; if it lasts for several hours and increases after subsequent efforts, they must be suspended. Kneading with one hand so as to break up indurations or disperse effusions, while the other hand pushes along the circulation in the veins and lymphatics above the joint is often a good procedure and quickly leads to the absorption of products not too firmly organized. Blisters and powerful derivatives in the neighborhood of joints affected with rheumatoid arthritis are considered more likely to promote than to retard the affection, according to the studies of Senator. Massage of the adjacent skin and muscles acts as a physiological derivative and raises nutrition to a high degree by a rapid interchange of materials, owing to the area and speed of the circulation being increased and obstructed lymphatics and capillaries made permeable. In this manner the soft structures may be made to adapt themselves to nodosities and deformities that cannot be removed.

When the disease is very active, or the muscles fattily degenerated, the tendons frayed out and thinned, and loose cartilages in the articular folds of the fibrous membrane, or when bony anchylosis has taken place, we would not expect anything from massage. All of my cases had been well nigh rubbed to death in the ordinary way, besides having exhausted the resources of the materia medica, baths and mineral springs of all kinds, electricity, etc., before coming to me for massage.

CASE I.—Miss A. was sent to me by Prof. Da Costa, of Philadelphia, in 1873. She was then 78 years of age, and had been bedridden for eight months with rheumatic gout. The disease had effected every joint of her limbs and apparently had run its course and done its worst. There was eburnation of the articular surfaces and rattling of the bones, with distortions of almost every joint. There was neither anchylosis nor contracture, but only slight indolent thickening around the joints with great muscular weakness. The patient was nervous and depressed. Pulse 92 and dicrotic. After one application of massage she could put her arms under the bedclothes and pull them up around her neck. Four days afterwards, having had daily massage, she could reach back so far as to get an object behind the pillow on which her head rested, sleep was better and she was greatly soothed by the manipulation. In seven days her pulse was down to 79 and steadier, and she walked across the room half supported. The following day she fed herself for the first time since her illness. In three weeks she could walk twice the length of her room with slight support. It seemed cruel to ask this patient to walk, so badly were her ankles deformed. But there was no pain; the sensitive structures of the joints were probably destroyed. This patient had massage 23 times in one month, and the use of her limbs and general condition were greatly improved. I fancy there may have been times during her illness when massage would not have been of much use to her. She was the most delicate of a family of six, having suffered from pleurisy, pneumonia, hemorrhages from the lungs, etc., and she outlived them all.

CASE II. — Mr. S., about fifty years of age, was sent to me by Dr. Thomas G. Morton, who had clearly explained to him the nature of his affection, and told him that benefit might result from a long course of massage. For a year the patient had suffered from periarticular thickening of the finger and wrist joints of a semi-solid nature, less of the elbows, ankles, and knees, in all of which places there were frequent but not severe attacks of pain of a non-inflammatory character. His general health had greatly deteriorated, sleep was poor, appetite small, and he had lost flesh, but still he could go out every day and attend to some business. Daily massage was used for a while, afterwards every other day. Improvement was slow and almost imperceptible. At the end of ten weeks, the periarticular indurations had decreased, the joints were more comfortable and more flexible, sleep and appetite had improved, and the patient had gained flesh. At the end of four months, massage was discontinued, the patient being to all intents and purposes well, having regained his natural sleep and appetite, and increased in weight to a marked degree, the circumferance of his wrists with the improved nutrition being the same as when the induration was present. There was still slight enlargement of some of the phalangeal joints, but the patient could shut his hands firmly, whereas at first they could only be semi-flexed. He said if he thought he could have been cured in this way, he would not have sold his house and business, with the expectation of going to Europe to seek for relief.

CASE III.—Mrs. A., G., thirty years of age, enjoyed good health until she was nineteen years old, when rheumatic gout appeared, first by swelling and pain in the last joint of the left thumb, and soon the other joints of both hands and fingers were similarly affected. Four months later the ankles became swollen. Then there was a respite for a year, and at the end of this time there was another exacerbation for five weeks. Six months later, it was apparent that the hands and feet were "growing out of shape." About this time she became pregnant, and the disease was in abeyance during this state, but after the child was born the joints were generally affected as before. Again there was a year of improvement, followed by grief and taking cold, which caused a relapse, and in 1877 she walked with great difficulty. Then iodide of potash, arsenic, a sojourn at Caledonia Springs, and electricity were tried, but the patient grew worse. In the summer of 1881, Dr. Goodrich, of Bakersfield, Vt., rubbed the knee-joints very hard, and hurt the patient a good deal, but she improved so that she could walk with a cane, and go up and down stairs, though the legs were greatly bent. She relapsed again in a few weeks.

This patient came to me in February, 1882, with knees flexed to almost a right-angle. There was slight effusion in them, and dry creaking on motion. There was thickening around the ankle-joints, though the motion was free and strong. Hip-joints not affected. The ring and little fingers of both hands were hopelessly deformed, the first row of phalanges being extended far back, the second and third sharply flexed and anchylosed. The thumbs, fore, and middle fingers were somewhat out of shape, but remarkably good use of them was retained. The left wrist was strongly flexed, the elbows and shoulders comparatively free, though they had been severely affected. The patient was bright and cheerful, and looked healthy, notwithstanding that she was suffering from general bronchitis, and weighed eighty-two pounds. Her appetite was fair and bowels constipated. A tonic pill to increase the appetite and regulate the bowels was given, and general massage used daily. Her joints were not very sensitive, and, from the first the massage was agreeable; the aches and pains disappeared in too remarkable a manner for me to ask anybody to believe. In three weeks she was eating very heartily, and taking five goblets of milk daily, and was free from discomfort; and after vigorous extension of the knees, which hurt at the time, they felt better all day, and we had gained one and one-quarter inches of extension at the ankle. In two months, the menses, which had been absent for eighteen months, returned, and they have continued to return ever since, with greater regularity than they ever did before. In three months, nutrition had improved to a marked degree. At the end of this time, the patient having been in preparation by a variety of impromptu, gradual extension apparatus, I put on each leg long lever splints, which made extension from the heel, and pried down the knees by means of wide bandages above them on the anterior aspects of the thighs, while the short arms of the lever rested on the back of the thighs.[1] These were kept on night and day for twelve days, and during the last seven days of their use the limbs made daily extension. After they had been on for twenty-four hours, the patient really liked them, and later, when temporarily removed, she felt faint for want of their support. They were then removed for eleven days, and light ham-splits applied, and with these the patient stood up. After six days more of the long splints, the legs were about as straight as people's usually are in use. By this time the joints had become sore, and the patient was not sleeping well. Two weeks later the patient walked twice across the room, pushing a chair before her to balance herself; ten days afterwards she did this twenty-two times, and in ten days more she walked with-

---

[1] I first saw this used by Dr. E. H. Bradford to straighten bent legs resulting from infantile paralysis.

out anything for support, except light wire ham-splints. Four weeks later she was going up and down stairs alone, and two weeks after this she frequently took a walk of half a mile with the aid of an arm and her ham-splints, then six months from the time treatment was begun, massage having been kept up all this time.   At the end of seven months and a half, the patient returned to her home, wearing the usual support for weak knees, but, unfortunately, when these were unfastened there was not much bend to the knees.   This, however, has improved by the patient's efforts.   The flexed wrist and movable finger-joints made good progress under massage.   Eight months after her return home the patient could sit down and get up from an ordinary chair, without arms, and run a sewing-machine.   Sixteen months later I removed entirely the knee-supports and gave her a short course of massage with increased improvement.   Eighteen months later the patient wrote to me that she had improved beyond all expectation ; had been out in all kinds of weather, the previous winter, and had enjoyed the novelty of walking upon the snow after having been confined to the house for nine consecutive winters.   When the thermometer had stood many degrees below zero, she had taken long walks, and while other ladies complained bitterly of the cold, she would be warm and enjoying the air hugely. Her legs had become straighter.   Nineteen months later she wrote to me again, saying that she found a continued increase of strength in the muscles of her legs, and could do housework all the forenoon without getting tired ; and a little after this she enjoyed climbing over snow-drifts.   Six months after this, in June, 1888, she came from her home in the country to me, with dyspeptic symptoms, but her walking powers were good.   The knee-joints had about two-thirds natural motion, and there was slight dry creaking, but no pain.   It really seemed as if new cartilage and synovial membrane must have formed, if such were possible.   It is remarkable that improvement should have continued, with so little treatment after her first course.

While I was first attending this patient, I read a report of a somewhat similar case treated by massage, with better results than could have been anticipated, and this inspired us with hope and perseverance.   It was the following :

CASE IV. — Dr. Carl Gussenbauer, Professor of Surgery at Prague, gave the massage himself in this case. The patient was a lady forty years of age, who had suffered from rheumatoid arthritis for twenty-two years. The trouble began in one elbow, and later affected the finger-joints. Rest and local applications were without effect. In the following years all the joints of the body were attacked. During two pregnancies no new joints became affected and those that had been were more endurable. The malady progressed in spite of medication, baths, mineral springs, electricity, etc. The patient had not walked any for eight years. Pains by day and night disturbed her rest, and she had become thin, anæmic, and constipated. The heart's action was irregular, and there were attacks of syncope at times. In this condition Prof. Gussenbauer undertook massage with great misgivings. He not only applied massage to the joints, but to the whole body — to the joints for the painful swellings that affected both upper and lower extremities, as well as some of the dorsal and lumbar vertebræ ; to the whole body to favor the nutrition of the atrophied muscles. There was but a small range of motion in any of the joints of the limbs, owing to adhesions in them, and in the sheaths of the tendons producing the characteristic contractions, and these he hoped to gradually overcome by passive motion. The deformity of the articular extremities of the bones, of course, could not be removed. He *masséed* the patient daily, and gave her passive and active motion, and after fourteen days he had the pleasure of seeing the swollen joints decreased in size, and the motion improved and less painful. At the end of a month the patient could walk about her room with a cane, and the improvement was so great that recovery seemed possible. For the next four months, the patient was *masséed* by the assistant, Dr. Schmidt. The result was that the pains and swellings of the joints and capsules disappeared, and they became movable to almost a maximal extent, except in one elbow, which had undergone bony anchylosis. The muscular strength of the patient had increased so that it often tired out the operator to give her the active or resistive movements. She could walk in her garden for fifteen minutes without fatigue. Nutrition improved more and more, the anæmia disappeared, the pulse became regular,' and menstruation returned, unaccompanied by fits of prostration as formerly. *Tapotement* or percussion with the ulnar border of the hand, with the palm, and with the fist was freely used in this case.

In several cases of chronic articular rheumatism, Prof. Gussenbauer reports favorable results from massage.

CASE V. — While some patients become thin under rheumatoid arthritis, others grow fat. Such was the case with Mrs. J. L., fifty-eight years of age, who had suffered with this disease for fifteen years. In seven months under massage, she regained free and natural use of her limbs, which she retained until a short time before her death. For the first two years her joints gave her but little trouble, but at the end of this time she was confined to bed for five months with arms so stiff and painful that when they were out of the bed-clothes she could not put them in again alone. Later the trouble left the arms and appeared in other places, but never in the toes. Her history is one of pains and stiffness of joints and muscles, with occasional remissions, the disease all the time slowly progressing. Her condition had been made much more comfortable by alkalies and laxatives prescribed by her physician, Dr. J. P. Oliver, who referred her to me for massage. Previous to her malady, she had been very thin and active, but for the last ten years of her illness she had grown very fat and heavy. For the last year before I saw her, her mode of locomotion indoors had been by means of a wheel-chair, and it was with difficulty she could be got in and out of a carriage. On each side of the patellæ it was evident through the layers of adipose that the capsules of the knee-joints were thickened, and it was to these places the pain of extension was referred, and this was limited to about three-fifths its normal range. Flexion was natural, and dry crepitus could be heard and felt on motion. Massage and movements every other day, with the use of curved ham-splints, and, later, straight ones, produced good results. In five weeks she was walking about the house with crutch and cane, and the legs could be straightened so that the popliteal spaces would touch an applied straight ham-splint, and free motion was gained to this extent. The splints proved a support and comfort in walking, and were worn for several weeks longer to overcome the tendency of the posterior femoral muscles to shorten. At the end of three months, the patient could walk the length of a room and back again without support of any kind, after getting started by assistance for a few steps. She had then lost the pain in the joints that troubled her by night, and pulling the legs no longer hurt the knees as formerly. When the knees ached, walking relieved them. The indurations and thickening were disappearing, and a hard, painful spot behind the left hip had entirely gone, and the patient could turn easily in bed. In six months the patient walked easily and gracefully without crutch, or cane, or splint. A month later, she got up from her chair alone and walked as well as any one of her weight could be expected to do, and the improvement continued for six months longer, until an affection of a different kind supervened, and suddenly carried her off. I need not say how sorry I was to lose this patient, who did me as much credit as any I ever had. The beneficial change in the knees was an indication of what had taken place in the other joints, in the fingers, wrists, and elbows, disappearance of indurations, and increase of motion.

CASE VI.— Miss E. P., forty-five years of age, at the change of life, well nourished, and a picture of health, has suffered from rheumatic gout and its variations for five years. During its early appearance, while crossing the Atlantic and returning, all the symptoms disappeared, only to return as soon as she landed. Two years after the disease began, she was seized with severe gastric disturbances for several months, and during the most of this time the joints were remarkably well. With the disappearance of the stomach trouble the mischief in the joints returned. The patient had derived, and still continues to derive, marked benefit from forty to eighty grains of salicylic acid given in divided doses during the day, with bi-carbonate of soda and carbonate of ammonia. This was prescribed by her physician, Dr. Lucy E. Sewall, who referred her to me for massage. As the patient was improving, I could see no need of adding this treatment, for cauterization of the knees was having a good effect, and making them feel better all day afterwards. It was agreed, however, to try massage on the worst limb, the right leg. There was a mass of feebly organized, indolent tissue in front of the condyles, which was easily dispersed, and its absorption hastened by massage, even the first sitting producing a marked effect, so that the adherent skin became free and the hyperplastic tissue softer. At the end of a week, with daily massage, this semi-indurated tissue had almost entirely disappeared, and the feeling as if the joint were held by an iron band in front of it had correspondingly lessened; the patient could stand more easily and push the limb against resistance more vigorously; the œdema of the leg and swelling of the ankle, which were partly controlled by a bandage, had also improved. During the first three weeks, daily gain was perceptible, and at the end of this time her physician asked me to apply the same treatment to the other leg, which had not been getting along so rapidly under cauterization as the worse one had been under massage. After this they improved *pari passu*. A splint was worn for twenty hours every other day on the right leg, and this was well tolerated and soon overcame a tendency to bending of the knee. The hands and arms made progress, and in two weeks she could use them more freely and write more easily. At the end of thirty-nine days, having had daily massage, the patient took a ride, getting in and out of a carriage with an ease that surprised herself. Three weeks before this, she could scarcely take a step, and for a year she had not been able to walk in any way worth mentioning. Massage was then used every other evening, and she always slept well after it, but not when it was omitted. Manipulation always made her lame and sore for several hours afterwards, only felt, however, on motion. She continued to gain under massage until she went to the country, two months from the time it was begun. She improved for three months while in the country, but on her return to the city she rather lost ground which massage scarcely made up for. After two months more of this treatment, it was suspended for two weeks, and during this interval she suffered from general stiffness, wakefulness, and fullness in the head, all of which disappeared promptly on resuming massage. And now, in spring-time, with the most unfavorable weather, and arduous professional duties, the patient is again improving. She has good use of her arms, and the thickening of the knee-joint capsules is decreasing, and she walks pretty well with very little assistance.

CASE VII.—Mrs. C. J., fifty-two years of age, has had rheumatoid arthritis for eight years, which has pursued a relentless course in spite of twenty-three physicians and all sorts of treatment. The joints of her hips and back are the only places free from the disease which is now pursuing its ravages in all the joints of the upper extremities, having for the present got through with the lower ones. Her general health is good, but before this trouble she was delicate and dyspeptic. Massage every other day for six weeks, though removing the periarticular swellings and thickenings and extending the knee-joints a little, which are almost bent at a right angle, has had no apparent effect except to make the joints lame and sore, and the patient tired. The upper extremities are almost anchylosed and resent passive motion. The patient perceives no improvement from the massage. This is the most discouraging case I have yet encountered. After the disease has spent itself on the upper extremities there will probably be a chance for massage to do some good.

By means such as we have been considering, Fuller, of London, has told us that, when the contracted tendons seem almost to require the surgeon's knife for their relief, operative interference will often be rendered needless. In one case, by persevering for nine months, he straightened a leg that had been flexed for four and a half years. Indications for encouragement he clearly expresses in these words : " When, as is usually the case, the contraction depends upon simple gluing together of the structures external to the joints, with the thickening of the capsule, and possibly some remains of effusion within the cavity, or into the bursæ external to the joint, I see no reason for despair as to straightening the limb, however long contraction may have existed."

CASE VIII.—Rheumatoid arthritis of a single joint resulting from injury is recognized. In one case of this kind where the knee-joint had been injured ten years before, and there was thickening, grating, slight deformity, and shortening, but free motion, though walking was exceedingly limited on account of pain, weakness, and the great weight of the patient. She had been rubbed an hour a day for a year by a very nice person, with the result of always being tired after the procedure, and getting no better. Twenty minutes' massage, with carefully increasing resistive movements every other day, for two weeks, improved her walking to an enjoyable extent. The induration was lessened, and the comfort of the joint increased.

A gentleman 63 yrs. of age had a fall five months before he came to see me on crutches with atrophy and relaxation of the quadriceps extensor from rheumatoid arthritis of the knee-joint, the result of his accident. There was slight subluxation of the head of the tibia backwards, some thickening of the capsule and periarticular tissues and dry crepitation on motion. He could not raise the leg extended either in standing or lying; but after three massages in one week he could elevate the whole limb extended at the knee for a second, and after five weeks' treatment he could hold the leg extended for 9 minutes in the horizontal position. The periarticular tissues had improved, the muscles had gained in size and firmness, but the joint surfaces had not improved, though he had got rid of "catches," and went without his crutches at the end of 17 days.

Father D., 27 years of age, always strong and wiry. When a boy had trigger-finger of the right middle finger. This disappeared when rheumatoid arthritis in the shoulder-joints began, 10 years before he came to me, and which was no doubt caused by straining the shoulders by holding himself too long on a cross-bar in the gymnasium. Both deltoids were atrophied and whipcord-like. He could dress and undress with difficulty, and there was much crepitation in the shoulder-joints. He could elevate the upper arms to about 75° from the perpendicular plane of the body, and on pulling back the left humerus its head slipped forward half way out of the glenoid cavity. He discovered sooner than I that massage and movements were of benefit to him and he kept them up for four months with the result that he could raise his humeri 115°

from the perpendicular. His deltoids had increased much in size and suppleness, and the shoulders were much more comfortable, so that his gesticulations as a clergyman had become easy to him. That no weight need be borne upon the shoulder-joints is perhaps a favorable circumstance for treatment in this condition.

No description of the effects of massage in rheumatic gout would be complete without reference to the case of old Admiral Henry. In 1810 he was seventy-nine years of age and had suffered from this affliction for twenty-eight years. There were swellings of his knees, ankles, and insteps which made him quite a cripple, and his fingers were also swollen and contracted. His stomach and bowels were hard, painful, and disordered. By means of various instruments made of bone polished smooth and hammers covered with cork and leather he persevered in the use of deep friction and percussion night and morning for three years. At the end of this time, it is said, he had completely succeeded in removing the swellings and had restored himself to the use of his limbs. His operations were at first painful, but ceased to be so after a little while, and soon they became so pleasant and useful that after having gone through with them in the morning he felt better all day. At the age of ninety-one he wrote to a friend: "I never was better, and at present am likely to continue so. I step up and down stairs with an ease that surprises myself. My digestion is excellent and every food agrees with me. I can walk three miles without stopping." This was in 1823, and he may be alive yet for anything I know.

The Admiral's pathology and therapeutics are of interest. He considered the chief cause of disease to be deficiency of circulation; and the best means of correcting this to prevent the nerves and muscles from falling asleep and getting fixed; for which purpose they should be kept loose by instruments worked among them. "By keeping the blood-vessels, nerves, and tendons in constant action by means of these instruments the blood is rendered pure, it passes quickly through the blood-vessels, leaving no *fur* behind it, so that ossification and stiffness are prevented." This quaint remark about the blood leaving no fur behind it is a noteworthy illustration of how a layman may express himself better than he is aware of, or in common language, "hit the nail on the head." This *fur* of old Admiral Henry corresponds to what we know of the white blood-corpuscles which move along so slowly in the circulation, having a tendency to adhere to the walls of the vessels, and which, when the current is retarded, as in points of inflammation, congregate in vast numbers and often migrate into the extra-vascular tissues, thence to be reabsorbed or become the foci of pathological changes.

In two cases of rheumatoid arthritis, Zabludowsky obtained marked improvement from massage, one of these, a recent case, apparently recovering. In two other cases of arthritis deformans of single joints from injury, benefit resulted from massage and passive motion to the extent of the patients being able to resume their occupations.

"The consequences of distortions and chronic rheumatic joint inflammations yield to the usual methods of treatment so slowly that one must be glad to have such a method as massage at his disposal, by which they come to an end comparatively quickly. Practice in the manipulations, time, perseverance, and personal interest in the matter are necessary." So says Prof. Billroth, of Vienna.

CHAPTER XVIII.

MASSAGE OF THE HEAD, FACE, EYES, EARS, AND THROAT.

ITS EFFECTS UPON MUSCULAR ASTHENOPIA, HYPERÆMIA OF THE RETINA, BLEPHAROSPASM, AND GLAUCOMA — UPON CHRONIC INFLAMMATORY PROCESSES OF THE ANTERIOR SEGMENT OF THE EYE AS EMPLOYED BY EUROPEAN OCULISTS, ETC.

Othello. — "I have a pain upon my forehead, here."
Desdemona. — "Let me but bind it hard, within this hour it will be well."
. . . .
Iago. — "My lord has fallen into an epilepsy; this is his second fit; he had one yesterday."
Cassio. — "Rub him about the temples."

To most people massage of the head is highly delightful, more agreeable indeed than on any other part of the body to which it is applicable, and in various disturbances as beneficial as it is pleasant. To account in great part for this increased comfortable sensation, we need only remember the acutely sensitive condition of the terminal filaments of the fifth pair of nerves, and that they will show signs of sensibility under circumstances to which spinal nerves would elicit no response. But massage of the head is seldom attempted, for manipulators are so accustomed to grasping muscular masses that, when they cannot do this, as on the skull, they are apt to think that nothing can be accomplished. Even Estradère is of the opinion that massage of the head is of little use, for no other reason evidently than that he is at a loss to know how it can be done. When manipulation of the head is attempted, it is usually, I might say almost always, in a way that would be better described by the word shampooing than by any other. Save with moisture for the purpose of cleanliness such a procedure had better be omitted, indeed, most people would object to such dry

309

rubbing on account of its "setting their nerves on edge." The soothing influence of gentle stroking, or of the hair being combed by another would be preferable. But the idiosyncrasies that cannot tolerate the following manner of doing massage upon the head are very rare.

With the patient in a semi-recumbent position in an easy chair that does not rock, the head inclined towards the side on which the operator sits or stands at a suitable distance to give his arms free play, one hand will be placed over the temporal muscle and fascia, and the other, if the manipulator be a novice, will be placed on the frontal region to steady the head, then the pressure of the hand on the temporal region is instinctively graduated to give the greatest movement of the scalp and underlying tissues between it and the bone, without gliding of the hand or such vigorous compression as hinders motion. Three or four manipulations will be made in this way and proceeded with step by step, the advance overlapping one-half of the region just worked upon, until the occiput is reached, and the whole can be repeated several times. After this the other hand can be trained to make similar movements in an opposite direction, proceeding from the superciliary ridges over the top of the head, upon the occipito-frontalis to the nape of the neck. A good *masseur* ought to be able to keep both hands going on these regions at the same time, one contracting as the other relaxes, without scraping, scuffing, shaking the head, or turning a hair. The manipulator having done one side of the head and the region bounded by the occipito-frontalis will find it most convenient to step to the other side of the patient and proceed as before. The back of the head can be more thoroughly *masséed* with the head erect, one hand steadying it on the frontal region, while the other makes manipulations in an oval direction, the long diameter being horizontal. At the middle of the upper portion of the back of the head, the centre from which the hair sheds off in all directions, a spiral circular movement with the palm of the hand is agreeable, efficacious, and useful.

Upon tough scalps that cling closely to the skull, massage is very hard work, and all the available motion that can be gained will often be by means of the ends of the fingers, and this only

of slight extent, proceeding in the same direction as when longer
sweeps can be made. In such cases, when the hair is long and
thick, by running the fingers through it close to the scalp, additional support will be afforded, which will secure more effectual
manipulation by their palmar surfaces, and prevent friction.   An
excellent way and pleasant to the patient to finish massage of the
head is to place one hand on each side of it, and make simultaneous
manipulations away from the median line from before backwards.
The groove between the occiput and back of the neck should receive special attention, by accurately adapting the palmar surface of the fingers to it as far as the median line, first on one
side and then on the other, and making upward and forward
manipulations.   Downward and backward manipulations have
but little effect here, though they may be used advantageously
with the thumb over the mastoid processes.

With such modifications as will readily suggest themselves,
the head can also be well manipulated while the patient is
lying down, and gentle stroking with light percussion can frequently be added with advantage.   The manipulator who fancies
he can sit behind a patient and find the "scalp loosely connected to the underlying skull" for his convenience of doing
massage in this disadvantageous manner, will find himself at
fault both in his position and supposition.

In *masséing* the face of a fat patient, the tissues can only be
rolled and stretched under the fingers and palm away from the
corners of the eyes and alæ of the nose toward the angle of the
lower jaw ; but if the patient be moderately nourished or thin,
the cheeks can be grasped between the thumb and fingers, and
more thoroughly squeezed and *masséed* in the same direction.
The eyelids can be effectually and agreeably manipulated by
pinching them up with a rolling grasp by means of the forefinger
and thumb at right angles to the orbicularis palpebrarum, and
stretching the enclosed fold away from the canthus to which it is
nearer.   With one thumb closely adapted to the inner portion
of the supraorbital arch and the middle finger of the other hand
a short distance below it on the upper and bony portion of the
nose, useful stretching in opposite directions and away from the

inner angle of the eye can be made, and should be simultaneous. Moving the thumb one step further along, so as to include the whole supraorbital arch, all the tissues upon and immediately beneath it can be advantageously stretched upwards and outwards, at the same time that the lower lid is pulled down by one or two fingers of the other hand by carefully graduated pressure upon the lower margin of the orbit. The alæ of the nose, one at a time, can be stretched and manipulated by covering the end of the finger introduced internally with a fold of soft handkerchief. By putting the index finger, still covered by a fold of soft cloth, inside of the cheeks, these can be squeezed, manipulated, and stretched between the thumb and finger, observing carefully not to lacerate the bridles of mucous membrane.

These manipulations seem easy enough to describe and to do, but nowhere in using massage are more practice and skill required than upon the head, face, and around the eyes.[1] Let any one practise them for a dozen years, and he will find that he can do them more easily and effectually the last year than he could the preceding one. The highly skillful and ingenious originator of lithotrity at a single sitting evidently takes more pleasure in telling how much better he can pass a catheter this year than he could a few years ago, than he does in demonstrating his valuable discovery of the toleration of the bladder and the manner in which he has taken advantage of it. Prof. Gross took more satisfaction in showing his pupils how to make a poultice than how to amputate at the hip-joint.

But what is the use of all this massage of the head? Its use and benefits may be as extensive as a morbid influence acting on the fifth pair of nerves is injurious. It is a plain saying, but a true one, that it is a poor rule that does not work both ways, and we see no reason why an agent like massage, that favorably influences the circulation, nutrition, and sensation of the branches of

---

[1] Since the publication of the first edition of this book I have been surprised to find that this method of *masséing* the head and face is not understood nor practiced outside of the city of Boston, except by some pupils whom I have instructed. Even in France and Germany the prevailing method is by stroking and percussion, much the same as the crude ways of the magnetizers.

the fifth pair accessible to its impression, should not manifest its benefits in as many different ways as an injurious influence does in other ways. Brown-Séquard enumerates no less than eleven different affections of the eyes that may arise from a supra- or infra-orbital neuralgia, in illustration of the general rule that " the same peripheric cause of irritation acting on the same centripetal nerve may produce the greatest variety of effects, including every functional nervous affection or disorder." ("Lectures on Functional Nervous Affections.") Besides the acute sensibility of the trigeminus, its influence over nutrition, taste, and smell, and by its anatomical relations it is also surmised over hearing, must not be lost sight of in considering the effects of any therapeutical agent applied to it. We have already referred to the effects of massage of the head in relieving peripheral neuralgias, and sometimes also those of central origin, and there is every reason to believe that deep-seated neuralgic pains in the periosteum, bones of the skull, dura mater, frontal sinuses, and orbits are likewise relieved by the sympathetic effects of massage, extending from the external branches of the fifth pair to those supplying the more deeply situated structures. A study of the effects of massage upon the head would extend, therefore to a study of the anatomy, physiology, and pathology of the fifth pair of nerves. We must content ourselves by indicating its results. These are not obtained by the random use of massage, for just as we have a sense of healthful and health-producing harmonies of sights, sounds, tastes, and odors, so also are we possessed of the faculty of perceiving agreeable sensations and motions conducive to our welfare, and nowhere upon the body where massage is applicable are its sensations and motions so quickly perceived to be in or out of harmony with the feelings of the patient as upon the head. Even when tissues are tough, matted and sore, here and elsewhere, patients of sense will often speak of the agreeable hurt of vigorous massage, and when this has ceased with returning suppleness, they may think that they are no longer being benefited. Aside from the effect of massage, perhaps no better illustration of healthful and agreeable impressions upon the fifth cranial nerve can be given than those arising from cool breezes on a hot day which impart to the brain a sensation of comfort, and excite in return reflex acts of breathing. The strength of massage, and the number of intermittent squeezes upon the head have much to do with the result; the former has already been indicated, the latter should be about sixty per minute with each hand for the majority of cases. If it be desirable to *masser* the head in giving general massage, ten minutes will suffice; but if the head require special massage, twenty or thirty minutes may be occupied.

The subjective effects of massage of the head are, in general, extreme comfort with a tendency to go to sleep, which, strange as it may seem, is equally consistent with an aptitude for mental work;[1] freedom of respiration through the nostrils, and light, clear feelings taking the place of dull, heavy ones. Increased suppleness of the scalp and tissues generally would seem to be the objective effect of massage, which often precedes improvement in detail of more important character, the most immediate and apparent instance of this being relief of congestion of the Schneiderian membrane and the ease of expelling tenacious mucus. The deep lymphatics of the face are derived from the pituitary membrane of the nose, and that these can be made more permeable by means of massage will account in great part for the resulting freedom of respiration through the nostrils when they have been obstructed and congested. Division of the fifth pair causes the nasal mucous membrane to swell and so disturbs its nutrition as to destroy the power of smell, the passages becoming obstructed by accumulated mucus. Pressure of effete matters upon the terminal branches of the fifth pair, would seem to have a similar influence, but to a less degree, and this pressure can be removed to a marked extent by means of massage. Then it need hardly be repeated that increased circulation in the external tissues of the head will do something toward relieving congestion in more deeply situated parts.

According to Pincus[2] idiopathic premature baldness is very common in men running from the vertex to the forehead between the parietal protuberances. Microscopically he has found an in-

---

[1] An eminent lawyer of this city always had his head rubbed before arguing a case in court.

[2] Berliner Klin. Wochensch. Nos. 4 and 5; 1875.

crease of the connective tissue of the cranium and a binding of this tissue to the layers beneath, which exerts destructive compression on the roots of the hair. Massage is certainly indicated in this condition, and ought to be beneficial if applied before the roots of the hair have been destroyed.

Besides the influence of massage of the head in relieving pain and headache of neuralgic, rheumatic, and sometimes of central origin, the most striking results I have obtained were in the relief of muscular asthenopia. My experience extends to but four cases, one an emmetropic, one a hypermetropic, one a myopic, and the last a myopic and astigmatic patient. They were cases of long standing, all in good health and not in any way run down, and had had their refraction attended to by the ablest ophthalmologists. Inasmuch as they had had good use of their eyes before, under apparently the same circumstances, why should they not again ? Massage of the head, and more especially around the temples, forehead, and eyelids, *but not upon the eyes*, produced marked and permanent improvement preceded by returning elasticity of tissues. I will only refer to the patient afflicted with myopia and astigmatism who came to see me in 1878, to ask me about trying massage. She was the worst of all four, and the one in whom the least improvement resulted. With concave cylindrical glasses, she had had good use of her eyes for many years until within the last six years, the trouble coming on from the time that she strained them "sight seeing" in Europe in side lights, poor lights, and all kinds of lights. I told her massage would probably relieve the strained, uncomfortable feelings, but that this would only be temporary unless she had appropriate spectacles. By the crude test of revolving each glass in front of the other eye than that to which it had been adapted, I found that better vision could be obtained, and sent her to an oculist without attempting massage. This physician told me that the case puzzled him, but he thought the trouble was mainly due to insufficiency of the internal recti. He gave her other glasses, treated her with prisms, and watched over the eyes carefully for four years, during which there was some improvement. But at the end of this time the patient decided to try massage, as she still suffered from tired, hot, uncomfortable feelings in her eyes with supraorbital neuralgia, lasting for several hours after every attempt to use them for fifteen minutes in sewing, reading, or fine work of any kind. After one application of massage, the eyes felt more comfortable and she could use them a little more easily. Of course this was but temporary, still it was a good omen. After four massages, tenderness and slight swelling over the left temporal region had disappeared. She had seven massages in one week in August, 1882, and in November she returned to me, for she felt that she had been somewhat benefited, though slightly. From November 15th to December 7th, she had nine massages and then at the end of this time she could use her eyes for an hour at a time in sewing, knitting, or reading with impunity. In December and January, her visits to me were few and far between, and there was no more gain. From the middle of February to the middle of March, she had twelve *séances*, sually going to the theatre after them and still, as might have been expected, there was no more apparent advance. She had several more massages with a little more encouragement — in all thirty-seven sittings. One year later I saw this patient, and she reported that her eyes had been much relieved from restraint and discomfort by the massage, that she had been able to use them longer and more easily, and that when hot, tired, and uncomfortable, they recuperated much faster and that this improvement was still continuing.

When this patient's eyes were comfortable before massage, they felt stirred up and uneasy for an hour or two afterwards, but after the perturbation had subsided they were easier than before the manipulation. If they were heated and tired before the massage they felt cool and comfortable after it, and this is usually the case. When massage was begun, her scalp was as tough, matted, and inelastic as sole leather, and it is not unlikely that this absence of suppleness extended to the areolar tissue and muscles of the orbit, especially to the internal and inferior recti. This state may have been a reflex effect of straining the eyes. Evidently its continuance must have had an unfavorable reaction in perpetuating the mischief, besides the disturbance it occasioned by its pressure upon terminal nerve-filaments and the hindrance to circulation, absorption, and nutrition. In such cases, and notably in this when the patient was asked to look upwards while the lower lid was being pulled down by pressure on the inferior margin of the orbit, she would make but a feeble effort, owing in all probability to stiffness of the tissues beneath the eye-ball and weakness of the muscles above; when asked to look downwards while the upper lid was being raised by pressure and extension on the upper margin of the orbit, the same feeble effort was observed, the eyes quickly rolling upwards, unable longer to bear the extension of the tissues above them. Such movements on the part of the patient, together with the counter-extension of the manipulation, are diagnostic and therapeutic at one and the same time, and I think clearly indicate that the condition of the tissues between the eye-ball and the walls of the orbit is similar to that found externally. In one of my cases of muscular asthenopia, however, the inability to look up or down against counter-extension of the lids was associated with almost natural suppleness of the external tissues of the head and face. This was the hypermetropic patient, an astronomer, and he improved so that he could see stars in the day-time (without falling on the side-walk) and read columns of newspapers. The practice of counter-extension for relaxing and elongating the tissues within the orbit is at first very disagreeable to the patient, but later becomes pleasant, restful, and refreshing to the eyes. It is to be expected, that the conjunctiva as it passes from the lids to the globe partakes in shortened growth with the other tissues. This must seem like curious pathology, but is it not consistent with similar conditions in other parts and with common-sense ? Upward and outward movements of the eyes in such cases cannot but be highly salutary in relieving the weakened and overstrained internal and inferior recti muscles, and I advised these patients to practise them several times daily. Turning the eyes in these directions tends to restore the harmonious distribution of will, nerve, and muscular action. Gymnastics with prisms to make the internal recti contract vigorously may be exceedingly beneficial, but I fail to see the philosophy of exercising muscles already over-taxed, at least until time for rest has been allowed. The conditions here seem to me to be quite analogous to those in writer's cramp and allied affections, over-use of any set of muscles giving rise to similar symptoms in those predisposed to them. Before resorting to division of the external rectus for the relief of muscular asthenopia, it might be well to try massage. An interesting form of passive motion is that recommended by Prof. Michel in the *Monatschrift für Augenheilkunde*, November, 1877, for ocular paralysis. This treatment, which succeeded in a case of rheumatic paralysis of the abducens, consisted in seizing the insertion of the affected muscle with a pair of forceps, and gently drawing the eye-ball as far as possible in the direction in which the muscle would move it, afterwards bringing it back again to its former position; this manœuvre being repeated for about two minutes every day.

But to return to the astigmatic patient. In seven days with seven massages, there was some improvement in the inert rigidity of the external tissues of the head and face, and long before massage was left off, the tissues of the scalp, face, and temporal region had improved in suppleness and elasticity to their utmost capacity, and the patient could look up and down to a normal extent while stretching was made upon the lids in an opposite direction. There was accompanying improvement in the use of the eyes, but not to a corresponding degree, though this improved after massage was over. [1]

Dr. Henry B. Stoddard tells me that he has treated successfully by massage of the head and face a case of muscular asthenopia of a year's duration in an emmetropic patient who had for a while been treated in the most approved manner without benefit.

Besides a few cases of chronic hyperæmia of the conjunctiva, that had been rather defiant of orthodox means, in which massage brought some relief, I have seldom ventured to try this remedy in ocular affections. Mr. W., suffering from hyperæmia of the retina, was sent to me in 1870, by Dr. Dyer, then of Philadelphia. Several years before he had had one eye removed, as it had been severely injured and gave rise to sympathetic ophthalmia in the

---

[1] Dr. Henry W. Williams, of Boston, expresses an unfavorable opinion upon tenotomy of the external recti in insufficiency of the internal, as it has not proved practically useful.

other. Massage of the head, face, and eye-lids proved an effectual revulsive and relieved the irritability of the eye to the patient's great satisfaction. The ophthalmoscopic appearances as observed by Dr. Dyer corresponded to the patient's feelings of improvement. The patient had frequently been troubled with hyperæmia of the retina before, but in his estimation he never got such prompt and complete relief as he did from massage. He was a gentleman of keen observation, in good health, and actively engaged in business.

In this case unnatural tension and toughness of tissues becoming supple was an accompaniment of improvement agreeing with the other signs. The tissues on the side of the head from which the eye had been enucleated were literally devitalized, tough, dry, tense, inert; on the other side less so. This brings me to say that the natural physiological action of any organ is essential to keep it and the adjacent tissues supple and healthy. So marked and constant is this that it can even be perceived by cultivated touch where there are only slight defects of sight and hearing and no other external symptoms of derangement. I like no better amusement amongst my medical friends than to feel of the tissues of their heads and faces, and in this way alone see how often I can ascertain which is the better eye or ear. I may with safety say that I have been able to judge correctly in at least eight cases out of every ten. In one case the difference in hearing between the two ears was so slight that the patient himself could not at first tell which had been affected many years before. In order not to mislead him or myself, I wrote my decision on a piece of paper, and after careful experiment the paper was read and found correct.

In a case of far advanced *glaucoma*, where one eye had already been operated upon, but too late, and it was not thought worth while to attempt iridectomy upon the other, I used massage every other day for four weeks, during which there was a stay in the course of the disagreeable proceedings, the neuralgia was relieved, and the patient made much more comfortable. It was impossible to keep up massage longer, and the unpleasant symptoms returned after it was left off, and hence we were led to think that it had been useful. This was was in 1876 (as the phy-

sician and relatives of the patient can testify), before Pagenstecher
and other foreign oculists had demonstrated the influence of
massage in diminishing intraocular pressure.

In the only two cases of distressing and long-continued involuntary twitching and contraction of the eye-lids in which I have
tried massage it proved successful. The mode of procedure was
massage of the head and manipulation and stretching of the
orbicularis. One was a military man whose nervous system was
somewhat run down. The fits of twitching disappeared after a
few massages, long before the tone of his system returned under
general massage. The other case I have just finished. It is that
of a lady, fifty years of age, in good health. For seven or eight
years she had been troubled with ;twitching of the lids, particularly the lower one of the right eye, and also the muscles of this
side of the face and occasionally also of the left hand and foot.
The involuntary movements were growing worse when she came
to me, and at times closed the lids. The twitching first came on
soon after severe mental trouble which still continues, and this
did not make 'the prognosis favorable. To make matters still
worse, she had been wakeful for four or five years, often not getting any sleep till four o'clock in the morning, and was sometimes
awake all night. Massage of the head and face with stretching
of the orbicularis and muscles of the face alone were used every
other day, and after four *séances* the involuntary movements were
scarcely perceptible for thirty-six hours. The effect not lasting
from one massage to the other, forty-eight hours, we tried manipulation every day for a while with better result, and, later, gradually lengthened the intervals to a whole week. The patient had
massage at three o'clock in the afternoon, and after five massages
she felt drowsy in the evening for the first time in her life. Sleep
soon became natural and easy, and the convulsive twitching almost entirely disappeared. Against loss of sleep and continued
mental anxiety this result is better than could have been expected.

Similar to these two cases, Dr. Ch. Abadie in *Gaz. des Hôpitaux*,
116, 1882, reports the following : A man, forty-five years old, had
suffered for about ten years from blepharospasm of the left eye

which had been treated without result. Neurotomy was proposed but declined. It was decided to try massage. The orbicularis was strongly stretched so that the skin and underlying tissues were acted upon. The sitting lasted six to seven minutes, the next on the second day ten minutes. In like manner the procedure was repeated for three weeks, at the end of which the lid of the affected eye obeyed the will as well as the other. In another case, a young man, the affection had lasted one year and he was cured in this way in two weeks. There was no result from massage upon an old woman who had suffered from intermittent spasm of both lids for a long time, so severely that during the cramp the lids could not be opened. Romberg states that direct pressure upon the facial nerve at its exit through the stylo-mastoid foramen will stop the blepharospasm in some instances.

Until quite recently the history of massage has presented but few instances of this agent having been used in affections of the eyes, and the results claimed have been so extraordinary that they will not gain easy credence without further confirmatory testimony by those best qualified to judge. M. Sabatier, physician and surgeon, in his "Traité Complet d'Anatomie" published at Paris in 1781, gives several observations analogous to the following: "Valsalva by his report confirms my conjecture, and I add it here to show that the ordinary resources of the art of healing were not used successfully in the following case. A woman was injured in one of her eyes by a turkey which she wished to capture. A little blood flowed from the wound. The sight of the eye was lost immediately. Many remedies were employed without relief. The patient then came to Valsalva who could perceive no external nor internal lesion of the eye. Reflecting upon this case, he imagined that by making vigorous frictions upon the course of the frontal nerves a beneficial change might be produced in the eye. Scarcely had he commenced them, when the patient recovered her sight. He attributed this success to cessation of spasm which he thought affected the muscles of the eye, and to the setting free of the optic nerve, which he supposed had been strangled." But it is more probable, Sabatier thinks, that it was due to the sudden shaking (*ébranlement*), molecular disturbance of the nerves of the eye. The reader may take this testimony for what it is worth. In those days it could certainly come from no higher source than Valsalva who was a pupil of Malpighi and later master of Morgagni. He was also professor of anatomy at the University of Bologna, which under his direction acquired great celebrity as a school of medicine.

Old Admiral Henry, whom we have already referred to, is a never-failing source of wonder to those interested in massage. In 1782 a cataract began to form in his left eye, but was unheeded until formed. Against the protestations of his friends, he was led to rub the eye-ball over the closed lids. He thought the eye was the better for it, and in hopes of dispersing the cataract he began perseveringly to use the round end of a glass vial. (It is a mercy it did not break.) Some time afterwards there was improved vision. He continued the practice, and in less than two years the cataract had disappeared and vision was again good in this eye. Two years later a cataract came in the other eye. It was operated upon by a distinguished oculist in London. Inflammation set in, and the eye was lost, but useful vision was still retained in the other eye. After the operation, the admiral suffered from excruciating facial neuralgia which extended to the eye that had been operated upon, and this reduced him to a great state of weakness. After having used various remedies with but temporary relief, he tried deep rubbing and thus got rid of his pains.[1]

I have heard of other cases of cataract disappearing under massage, but my enthusiasm has not yet been sufficiently aroused to try such means. While waiting for the necessity of an operation, massage in such a case would be a harmless and luxurious experiment, and if the patient could afford it and be responsible for the want of result, it might be justifiable. The instances in which cataracts improve for a time, or even disappear of their own accord, are so exceedingly rare that improvement or recovery following massage would be just ground for encouragement from this source. Spontaneous recoveries are generally indications that nature can be artificially aided.

---

[1] The Anatriptic Art, by Walter Johnson, M. B.

Chronic conjunctivitis, specks, and opacities of the cornea of small extent are said to have been cured with massage not directly applied to the eye-ball by Neumann (Roth, 1851). It is to be regretted that Neumann was not an oculist, and that he did not flourish in these days, for his experience has of late been confirmed by trustworthy ophthalmologists. These most careful, skilful, and scientific specialists have been the last to avail themselves of the use of massage. Its lists of advocates in affections of the eyes is headed by the most distinguished of them all, Prof. Donders, then follow Prof. Mauthner, Drs. Pagenstecher, Klein, Just, Schmid, Rimpler, Pedraglia, Schenkl, Bull, Becker, Damalix, and others. Prof. Leber, in his prize essay of 1873, showed that the ways of resorption and flowing off of the intraocular fluids converge near the corneo-sclerotic margin, another instance, one might be disposed to think, that Providence had planned the eye, like many other structures, to be subject to the influence of massage. It has taken humanity a *mighty* long time to find it out. Prof. Mauthner is of opinion that there is no organ so appropriate for the use of massage as the eye.[1] From numerous experiments and preparations, Dr. Leopold Weiss, Docent at the University of Heidelberg, has demonstrated that there are four ways for the flowing off of the intraocular fluids running to the corneo-scleral margin: 1, by means of that first described by Knies which runs backwards to the outer surface of the sclera from the ligamentum pectinatum and interior of the sclera; 2, by means of the vessels that penetrate the walls of the bulb at this place; 3, along the course of the strong connective-tissue fibres which can be seen running from within outwards; 4, from the anterior chamber into the cornea, and, out of this, opposite the margin of the cornea into the subconjunctival tissue. (*Centralb. f. pr. Augenheilkunde*, 1880, p. 87.)

At the Ophthalmological Congress, held in London in 1872, Prof. Donders called attention to the practice of massage as one that had given him excellent results in abscess of the cornea. (The last affection in the world we would have supposed massage to be applicable to.) The experience of Donders has been confirmed by Dr. Osio (*Independencia Medica* of Barcelona), and by Dr. Just

[1] Glaucoma, Wiesbaden, 1882: Bergmann.

(*Centralbl. fr. pr. Augenheilkunde*, June, 1881). Just observed that a speedy and easy absorption of hypopyon occurred when the pus was spread over as great a space as possible by means of a pressure-bandage. For this reason he attempted massage of the eyes by making gentle upward stroking and rubbing motion by means of a finger placed on the lower lid in order to disperse the pus over the whole of the anterior chamber. In this manner its rapid removal was obtained. By way of illustration he reports the case of a woman, sixty-four years of age, afflicted with *ulcus corneæ serpens*, *hypopyon*, and *blennorrhœa* of the lachrymal sac of three weeks' duration, to whom he applied massage night and morning and after the fifth sitting the improvement had been so great and rapid that iridectomy for exudation into the pupil was not considered necessary, the ulcer had become perfectly clean, the cornea much clearer, and the purulent matter had entirely disappeared and did not re-accumulate. Such speedy improvement Just believes cannot be obtained in any other way, and he advises others to try the procedure.

Dr. Pagenstecher, of Wiesbaden, seems to have been the first ophthalmologist to use massage upon the eye-ball after Donders had recommended it, but apparently independently of this recommendation. In one case that for twenty years had been subject to periodical attacks of inflammation of the eyes, presenting an affection of the whole sclera with uniform conjunctival and subconjunctival injection to a considerable degree, the conjunctiva of the globe slightly œdematous, with here and there at the margin of the cornea, small, round or oval prominences, not unlike the formations which are described as cystoid cicatrices, by the use of massage he succeeded in suppressing the attacks in their first stages. Immediately after the massage, diminution of intraocular pressure was always observed in this and other cases. Pagenstecher also used massage in a case of episcleritis, and in another of parenchymatous keratitis of specific origin. The episcleritis, in the course of ten days, was completely removed, having lasted for four weeks, and he never observed the like with other methods of treatment. The keratitis was in the stage of retrogression, but for a long time had been at a standstill. After the use of massage for a few days, a marked improvement in the power of vision was obtained. In these cases intraocular pressure was less after massage. His method of using massage was by moving the lids under slight pressure in a radial direction from the centre of the cornea, as quickly as possible; and after this by making circular friction under slight pressure upon the the upper lid around and upon the region of the sclero-corneal margin. His idea was that massage might succeed in removing hindrance to the circulation, and in this he was not disappointed, as it emptied the blood-vessels and lymphatics at the sclero-corneal margin and thus promoted rapid absorption of exudation around them. This was Pagenstecher's early experience with massage which was published in 1878 (*Centralbl. fr. pr. Augenheilkunde*, Dec.). Three years later, he gives his more extended experience and results in the use of massage, and these are still very favorable. [1]

In brief he found this treatment useful in chronic inflammatory processes of the anterior segment of the eye ; contra-indicated when it is found to cause excessive injection, and especially if there be photophobia and lachrymation ; and not to be employed in the presence of iritis. In some cases it acted as a depletive, lessening the tension of the eye ; in others as a stimulant, aiding the formation of new vessels, and thus proving beneficial when nutrition and absorption were torpid. Attention is called to its stimulating effects on the vaso-motor nerves, whereby a better contraction of the vessels resulted, though the immediate effect seemed to be that of dilatation, for after massage the eyes are still more injected, but on the following day this is less than before manipulation. The irritation produced must be of moderate degree, and should wholly disappear in half an hour. He used massage once a day from two to four minutes, and sometimes twice daily when it was well borne. Affections of the cornea, conjunctiva, sclera, and ciliary body were those in which he found it applicable, namely :

1.   Opacities of the cornea resulting from pannous keratitis, scrofulous superficial keratitis, and even parenchymatous keratitis. When, after corneal opacity has subsided, such opacities remain

---

[1] Arch. of Ophth., Dec., 1881.

stationary, massage re-excites a moderate vascularity, and promotes removal of the opacity.

2. Chronic pustular conjunctivitis, especially in old people.
Also in forms of chronic conjunctivitis in which there is a hyper-
trophic thickening of the membrane close to the margin of the
cornea, occurring either as an elevated yellowish wall surrounding
the cornea, or as one or more thick vascular papules, toward which
large veins course from the conjunctiva. A form of conjunctivitis,
chiefly caused by external irritation, in which the inflammation
occurs in a triangle with its base at the outer, rarely at the inner
margin of the cornea, the membrane being swollen and of a gray-
ish-yellow tinge and the conjunctival and the sub-conjunctival
vessels swollen.

3. Forms of scleritis and episcleritis in which fixed nodules
appear in or on the sclera, often accompanied with severe ciliary
neuralgia. Constitutional treatment is required in addition to the
massage, and the latter is not employed if there be iritis. Massage
appears to hasten the absorption of the nodule. Chronic episcleral
inflammation without iritis leading after long periods to alterations
in the tissue of the sclera.

4. Circumscribed affections of the ciliary body. In the one
case thus treated, a localized congestion of long standing in the
upper part of the ciliary region, associated with extreme sensitiveness and pain after efforts of accommodation, was cured by
massage.

Pagenstecher often uses a small quantity of yellow precipitate ointment under the lids, which makes them glide more easily
in doing massage, besides distributing and dividing the ointment
in the finest possible manner, so that its specific effect is greatly
developed. The experience of Pagenstecher has been confirmed
by several of the oculists mentioned, much to their agreeable
surprise. He obtained the prettiest results in cloudiness of the
cornea, whether superficial or deep-seated, partial or extending
over its whole surface. He affirms that the usual means employed
of clearing up the cornea, such as moist warmth, yellow precipitate, calomel, tincture of opium, etc., are far behind massage in
celerity of action.

Dr. Pedraglia, of Hamburg, used massage and nothing else in two quite recent cases of episcleritis, in which the exterior parts of the bulb presented well-marked circumscribed redness of the episclera without any trace of pustules or affection of the cornea. Massage was used every other day, and after six times one case was discharged well, and after fourteen days the other was considered cured. In the one case after four, and in the other after three massages the congestion had almost entirely disappeared and the disagreeable feelings of pressure had ceased (*Centralb. f. pr. Augenheilkunde*, April, 1881).

Panas used massage, and found it speedy, agreeable, and effectual, not painful, and rendered the best service together with precipitate ointment in chronic affections of the cornea in young people. In his hands it proved particularly effectual in parenchymatous and scrofulous keratitis and *pannus granulosus* (*Centralb. f. pr. Augenh.*, 1881).

Gradenigo states that a patient of his who in the evening required narcotic injections, entirely replaced these with advantage by massage of the eyes. Diminution of tension resulted, and pain and disturbance of circulation ceased. He also states that a decrease of tension occurs in healthy eyes from massage applied to them for two or four minutes, and at times pain will be felt, but this disappears and is followed by agreeable feelings (the same journal, 1880, p. 123).

Friedmann found massage of no use in *conjuntivitis phlyctenulosa vesiculosa*, but, on the contrary, it quickly cured *conjunctivitis phlyctenulosa miliaris*, and by using a little vaseline under the lids, he found that cloudiness of the cornea cleared up quickly (*Wien. Med. Presse*, 1882, No. 23).

Schenkl confirms the observations of Pagenstecher. He obtained the most favorable results from massage in opacities of the cornea in *keratitis profunda*, as well as pannous and maculous opacities of the most divers durations. He tried it in acute inflammatory processes, recent parenchymatous keratitis, in relapse from iritis, and in ulcer of the cornea, and found that it was well tolerated, but did not shorten the durations of these affections. In various forms of glaucoma, the good effects of

massage were only transitory, the diminution of intraocular pressure not lasting more than twenty-four hours. Secondary glaucoma was, however, an exception, for in such cases massage produced permanent improvement. Favorable results were obtained
in hemorrhage into the anterior chamber and into the conjunctiva. He also used massage in other affections of the eyes, but the
results are not stated (*Prager Med. Wochenschrift*, 1882).

In the service of Prof. C. E. Michel was a patient whose left
cornea was hazy and there were three large flat nodules over
(probably under) the central part of the tarsal cartilage. The
electro-cautery being out of order, massage with yellow precipitate
ointment was used instead. After two *séances* weekly for three
weeks the nodules had decreased in size so the cautery was no
longer thought of, and massage was persevered in. The cornea
became clear and the elevations scarcely perceptible. Another
patient of Prof. Michel's had trachoma for several years. This
had almost disappeared, but the uppermost portion of the cornea
was occupied by a mass of blood-vessels and the corneal tissue
was infiltrated and elevated. Massage with ointment of yellow
precipitate was used, and the cornea became perfectly smooth and
almost as clear as the well one.

In 1876, Max Knies published his discovery that the angle of
the anterior chamber is almost always found closed in glaucomatous eyes, and this fact became worthy of great consideration,
in view of the previous observations of Prof. Leber that the
aqueous humor in the normal eye here finds its way for flowing
off. Knies thought that the occlusion was owing to a primary
inflammation, and in this he was confirmed by Adolph Weber
(*Centralb. f. pr. Augenheilk*). Prof. Mauthner is of the opinion
that, in the first stage of glaucoma, its dangerous progress may
be arrested by means of massage ("Glaucom," Wiesbaden, 1882).
How vastly important it is even to arrest the progress of such a
dangerous affection! I have previously stated that it is in the
early and late stages of many affections that massage is most useful; in the early as a preventive, in the late as—well, that depends on the amount of injury done.

Dr. Karl Grossman reports favorably on massage in the interior of the eye in glaucoma. After paracentesis and the escape of the aqueous humor, he takes a club-ended silver probe, previously shaped at the end like a button-hook, and passes it into the anterior chamber through the corneal wound; the convexity of the hook being directed to the ciliary region, he tries to push it gently but decidedly forward between the cornea and iris as far and in as large a circumference as possible. With the hook, he gently presses the peripheral part of the iris back towards the lens, where a distinct resistance can be felt. This proceeding he repeats several times, and having done so in one quadrant, he turns the convexity of the probe round and goes to another. He is of the opinion that the improvement in three cases, one being desperate, in which he tried this, must be attributed to a successful mechanical re-opening of the iris-angle, which, once established, became permanent. Fears as to reaction and irritation were not realized. The operation eases the obstructed angle better than iridectomy, and avoids the drawback of the latter, viz., deformity of the pupil (*Ophthalmological Review*, London, 1882).

As a means of hastening resorption of the lens after *discisio cataractæ*, Junge, Chodin, and Becker testify to the favorable influence of massage. According to Prof. Junge, who had used it for this purpose for four years prior to 1880, it is effectual in two ways: 1st, upon the region of the sclerotic, the substance of the lens in the anterior chamber is pressed out of the capsule; 2d, upon the region of the cornea, the substance of the lens is finely divided and made more fit for resorption. Only when all appearances of irritation have ceased does he make use of massage for one or two minutes every two or three days (*Centralb. f. pr. Augenheilk.*, 1880, p. 279).

Dr. Van Der Laan has used massage of the eyes for traumatic cataract. A five-year-old child received a linear wound of the cornea and iris, and rubbed the eye frequently itself, with the result that, in twenty-four hours, the glaucomatous symptoms had disappeared and a great part of the swollen lens was removed from the pupil into the anterior chamber, from whence it was resorbed (*Ibid.*, 1881, p. 449). In this case, permission to rub the eyes was probably neither asked nor granted. A person falls and injures a joint, resulting in its becoming stiff. Again, the same person may fall, break the adhesions, and cure the joint; but they are quite as likely to break a limb, and so it would be if massage for affections of the eyes or any other organs were intrusted to the uneducated. But the teachings of Nature are not to be neglected, for instinct is a great matter. Every time we wink, the exterior of our eye-balls and the interior of the lids receive a sort of gentle massage conducive to their welfare. After heavy slumber, many instinctively overcome the inactivity of the lids by rubbing them, and it is a common practice to do this in order to see more clearly and to free the eyes from secretions.

Reibmayr points out that the reflex effects of massage can be studied upon one's self. He mentions the following uniformity of results : With both eyes closed, the one that is not being *masséed* will show a decidedly greater enlargement of the pupil than the other within the first minute. But with the continuance of the massage, the pupil of the same eye that is not *masséed* becomes smaller than the other in the second minute; at the end of the third minute, the pupil of the *masséed* eye is decidedly smaller than the other, both always examined simultaneously.

Klein is an earnest advocate for the use of massage in certain affections of the eyes. He makes no appeal from the experience of others. He has used it in episcleritis, phlyctenular ophthalmia, diffuse parenchymatous keratitis in the acute stage, and in phlyctenular and granular ophthalmia. He believes massage would be of use in glaucoma where it is not a question of operation, in hydrophthalmia, in ciliary neuralgia, in supra-orbital neuralgia, and in idiopathic blepharospasm.

"It is easy to understand," says Klein, "what massage effects when we reflect how many cases of chronic inflammation there are that leave behind them relics that are almost immovable ; thanks to its intervention, one is in much better condition to break up artificially the vicious circle. An inflammation with an abundant exudation provokes a deposit of new products, which on their part become the cause of a new inflammatory irritation and the formation of other deposits. All these will disappear as soon as the obstructed stomata become permeable. Most of the methods employed attain this view in one case or another, but in the majority of those of which it is here a question they fail. How should we proceed in the case of pannus become permanent? If the nutritive and productive source, conjunctival granulations, exists, it is to these that the therapeutics will be directed. The ordinary means are absolutely without action against pannus when there are no conjunctival granulations. The insufflations of calomel, the instillations of divers liquids are no other than irritants. What use are excitations of all kinds if the preformed products cannot find a way out? We incise part of the conjunctiva to make the vessels disappear which nourish the pannus, and new vascular connections establish themselves very quickly.

"The inoculation of pus alone answers the physiological postulate; it produces new ways of absorption and sets free the old ones. But at the same time it determines a veritable resolution in the structure of the eye; it always puts it in peril, sometimes destroys it. With massage it is quite another thing; here it is no more than a simple irritant, but capable of re-establishing the permeability of obstructed ways, of breaking up the pernicious progress. It is without peril, efficacious and mild, yet active and well supported by the patient. With all other procedures a good deal of time is required in successful cases; massage shows its favorable effects very quickly." (*Wien. Med. Presse*, 1882.)

Prof. Hirschberg, reports the case of a female, 26 years of age, who suffered from insufficiency of the mitral valves. The variations in the filling and emptying of the blood-vessels of the *fundus oculi* could be accurately seen with the aid of the ophthalmoscope. Massage of the bulb by pressing this against the orbit was used from the day after the blindness, and was repeated twice daily. On the second day sight returned in one part of the field of vision, and on the third day this was more extended, and was confirmed by ophthalmoscopic observation so far that the vascularization of the retina appeared everywhere almost normal. Yet

PRACTICAL TREATISE ON MASSAGE.

the field of vision was limited, the arteries attenuated, the papilla pale.[1]

To have the effects of massage upon other parts of the body which are hidden from view, seen, observed, and confirmed in the structures of the eye by oculists of the highest eminence is more pleasing and satisfactory than words can express. I venture the prediction that in all probability the best results will yet be obtained in affections of the eyes appropriate for massage by combining its local application with that of the head and face, whereby a much greater influence will be exerted upon the fifth pair, vaso-motor and sympathetic nerves, and a better revulsive effect obtained than by local massage alone.

In otology less than in any other branch of medicine has massage been found necessary. After the healing of the incision for the evacuation of the contents of *othæmatomata* in two cases, Dr. Clarence J. Blake sent them to me for massage. With a very few applications of this the slight puckering and induration of the auricle disappeared and the tissues became supple and of natural appearance. Compression had been previously used by Dr. Blake, and there seemed but little left for massage to do. Perhaps I under-estimate the importance of the procedure, for Meyer (*Archiv für Ohrenheilkunde*, XVI., p. 161) says that no other method of treatment can prevent the resulting disfigurement of the auricle. He mentions three cases in which, after the incision had healed, massage and compression restored the auricle to its normal condition and appearance in one week. Politzer recommends massage and centripetal stroking over the region of the mastoid process, in front of the ear and upon the side of the neck, as a means of relieving the pain of *otitis externa* and furuncles of the external auditory canal, especially when other procedures cannot well be used; but incision and local applications should not be neglected. Massage in these cases has a powerful depletive effect, diminishing the redness and swelling of the mucous membranes in the narrow canals of the auditory apparatus. The antiphlogistic effect of centripetal stroking of the neck has proved very efficacious in in-

---

[1] Centralblatt für Augenheilkunde, Vol. viii, 1883.

flammatory conditions of the Eustachian tube and middle ear, according to Gerst.[1] Massage is said to have been very beneficial in "nervous deafness," whatever that may be (probably an affection somewhat like amaurosis before the ophthalmoscope was discovered — "a disease in which patient and physician were blind").

Hommel[2] discards the usual modes of mechanical treatment of the membrana tympani and ossicles, and advocates a new method which he styles "*Tragus-Press.*" Acting upon the supposition that pressure of the tragus upon the meatus renders the canal airtight and condenses the air therein sufficiently to act directly upon the membrana tympani, he releases the pressure upon the tragus, and the sudden escape of the compressed air produces traction upon the membrane. Compression and attraction of the membrane are thus rapidly produced by exerting and relaxing pressure upon the tragus. Hommel suffered from chronic catarrh of the middle ear, and experimented in this way upon himself, and found that his hearing improved within several months from 10 and 40 cm. from the watch to 150 and 610 cm. A similar result was obtained in a boy 13 years of age, who suffered from a perforation on one side, and thickening and retraction of the membrane on the other. Prior to this mode of treatment the watch was heard only at 5 cm. by the left ear and 12 by the right. After nine months the hearing power had improved to 160 and 340 cm. respectively. Hommel advises this mode of treatment in (1) chronic catarrh of the middle ear; (2) perforations of the membrane, the handle of the hammer still being in connection with the membrane; (3) in thickening and opacities of the membrane; (4) as a prophylactic measure against deafness developing in old age. For this purpose a much more agreeable and efficacious method than that of Hommel is to apply the hand so that the muscles in front of the metacarpal bone of the thumb will fit accurately into and over the auricle; the right hand for the left ear, the left for the right, and by making moderately strong circular kneading upwards, backwards and outwards, powerful but pleasant compression and suction are created

---

[1] Gerst: Massage. Wurzburg, 1879.
[2] Archiv für Ohrenheilkunde, Bd. xxiii. p. 17.

which have none of the disagreeable feeling of shooting peas into the ear that the "tragus press" of Hommel has. Indeed Hommel's method makes no suction at all, for the compression of the air that it causes is so slight that its sudden liberation cannot produce any appreciable vacuum, as claimed by its advocates. When desired the auricle also can be grasped at the same time by the fingers behind it, and a circular traction with the strongest pull upwards, backwards and outwards exerted; but this does not afford so much compression and suction as when the auricle is not grasped. These procedures can be advantageously varied with *effleurage* of the neck so as to rapidly aid the return of blood and lymph, and thus deplete not only the ear but the whole head as well. The head and face may also be manipulated, which besides its revulsive effect will also have a sedative influence upon the ear by means of the communication of the fifth pair of nerves with the otic ganglion.

In catarrhal affections of the *nose*, of the *pharynx* and of the *larynx*, Gerst has used centripetal *effleurage* of the exterior surface of the neck, which exerts an aspiratory force, here as elsewhere, in depleting the veins and lymphatics, and the results have exceeded his expectations.[1] To 1879 he had treated in this way:

21 cases of acute catarrh of the pharynx;

10 cases of naso-pharyngeal catarrh;

9 cases of catarrh of the larnyx;

2 cases of catarrh of the larnyx and pharynx in consequence of syphilis;

1 case of catarrh of the nose and of the pharynx with ulcerations of the nasal mucous membrane (*ozœna syphilitica*);

1 case of chronic catarrh and ulceration of the larynx in a patient with phthisis.

In all of these cases but the last one, Gerst's anticipations were "brilliantly fulfilled." In the acute cases, even after a single sitting, the following improvement was observed: decrease of redness and tumefaction of the mucous membrane of the eyes, nose, and throat; disappearance of feelings of heat and

---

[1] Loco citato.

pressure ; respiration easier ; burning sensation in the throat in catarrh of the larynx lessened and voice clearer ; dysphagia and sensation of pressure diminished in pharyngeal catarrh, and the symptoms of hyperæmia of the brain and congestion of the frontal sinuses were also markedly lessened. Recovery ensued from repeated massage in astonishingly short spaces of time : on the average in the cases ₗof naso-pharyngeal catarrh in 3.2 days ; in those of pharyngeal catarrh, 1.8 days ; in those of laryngeal catarrh, in 3.5 days. In the two cases of *angina* and *laryngitis syphilitica* which had been previously treated with mercurials, recovery took place in these local symptoms, in one case in five days, and with the other in seven days ; and in the one case of *ozœna syphilitica* which had previously been medicated locally and constitutionally, recovery is said to have oc. curred in twelve days.

We cannot think that a military surgeon of such high standing as Dr. Gerst wishes to convey the impression that he would not use specific remedies where indicated, before, after, or with the massage. His intention is doubtless to show the effects of massage upon local processes of an inflammatory nature.

" And chief the illustrious race, whose drops and pills
Have potent powers to vanquish human ills ;
These with their cures a constant aid remain
To bless the pale composer's fertile brain ;
Fertile it is, but still the noblest soil
Requires some pause, some intervals of toil ;
And they at last a certain ease obtain
From Katerfelto's skill, and Graham's glowing strain."
                                                      —*Crabbe.*

INDEX.